Urban Sanitation Planning Manual
Based on the Jakarta Case Study

APPROPRIATE TECHNOLOGY FOR WATER SUPPLY AND SANITATION

Volume 14

In 1976 the World Bank undertook a research project on appropriate technology for water supply and waste disposal in developing countries. Emphasis was directed toward sanitation and reclamation technologies, particularly as they are affected by water service levels and by the ability and willingness to pay on the part of the project beneficiaries. In addition to the technical and economic factors, assessments were made of environmental, public health, institutional, and social constraints. The findings of the Bank research project and other parallel research activities in the field of low-cost water supply and sanitation are presented in this series. Other volumes are:

1. Technical and Economic Options [condensed from Appropriate Sanitation Alternatives: A Technical and Economic Appraisal, Johns Hopkins University Press, 1982]

1a. A Summary of Technical and Economic Options

2. A Planner's Guide [condensed from Appropriate Sanitation Alternatives: A Planning and Design Manual, Johns Hopkins University Press, 1982]

3. Health Aspects of Excreta and Sullage Management - A State-of-the-Art Review [condensed from Sanitation and Disease: Health Aspects of Excreta and Wastewater Management, John Wiley and Sons, 1983]

4. Low-cost Technology Options for Sanitation - A State-of-the-Art Review and Annotated Bibliography [available as a joint publication from the International Development Research Centre, Ottawa, Ontario, Canada]

5. Sociocultural Aspects of Water Supply and Excreta Disposal

6. Country Studies in Sanitation Alternatives

7. Alternative Sanitation Technologies for Urban Areas in Africa

8. Seven Case Studies of Rural and Urban Fringe Areas in Latin America

9. Design of Low-cost Water Distribution Systems

10. Night-soil Composting

11. Sanitation Field Manual

12. Low-cost Water Distribution - A Field Manual

13. Meeting the Needs of the Poor for Water Supply and Waste Disposal

Additional volumes and occasional papers will be published as on-going research is completed. Except for volume 4, all reports may be obtained from the Publications Sales Unit of the World Bank.

WORLD BANK TECHNICAL PAPER NUMBER 18

Urban Sanitation Planning Manual
Based on the Jakarta Case Study

Vincent Zajac,
Susanto Mertodiningrat,
H. Soewasti Susanto
and Harvey F. Ludwig

The World Bank
Washington, D.C., U.S.A.

This is a document published informally by the World Bank. In order that the information contained in it can be presented with the least possible delay, the typescript has not been prepared in accordance with the procedures appropriate to formal printed texts, and the World Bank accepts no responsibility for errors. The publication is supplied at a token charge to defray part of the cost of manufacture and distribution.

The views and interpretations in this document are those of the author(s) and should not be attributed to the World Bank, to its affiliated organizations, or to any individual acting on their behalf. Any maps used have been prepared solely for the convenience of the readers; the denominations used and the boundaries shown do not imply, on the part of the World Bank and its affiliates, any judgment on the legal status of any territory or any endorsement or acceptance of such boundaries.

The full range of World Bank publications, both free and for sale, is described in the *Catalog of Publications*; the continuing research program is outlined in *Abstracts of Current Studies*. Both booklets are updated annually; the most recent edition of each is available without charge from the Publications Sales Unit, Department T, The World Bank, 1818 H Street, N.W., Washington, D.C. 20433, U.S.A., or from the European Office of the Bank, 66 avenue d'Iéna, 75116 Paris, France.

Vincent Zajac is with Motor-Colombus, Switzerland; Susanto Mertodiningrat is assistant to the general director of the Ministry of Public Works, Jakarta; H. Soewasti Susanto is head of the Health Ecology Research Centre of the Ministry of Health, Jakarta; and Harvey F. Ludwig is a consultant to the World Bank.

Library of Congress Cataloging in Publication Data
Main entry under title:

Urban sanitation planning manual based on the Jakarta
   case study.

   (World Bank technical papers ; no. 18. Appropriate
technology for water supply and sanitation ; v. 14)
   Bibliography: p.
   1. Sanitary engineering--Developing countries.
2. Jakarta (Indonesia)--Water-supply. 3. Jakarta
(Indonesia)--Sewerage. I. Zajac, V. II. Series: World
Bank technical paper ; no. 18. III. Series: World Bank
technical paper. Appropriate technology for water
supply and sanitation ; v. 14.

TD353.U7 1984   363.6'1'095982   83-14817
ISBN 0-8213-0249-3

ABSTRACT

The provision of affordable water supply and sanitation services to all population groups--rich and poor--in urban areas requires the use of a variety of technologies, supported by information and education activities. Experience has shown that beneficiaries, in particular those living in areas with few or none of the customary municipal services, need to understand the purpose, the cost and the operation of the proposed improvement if they are to enjoy the intended health and economic benefits. As a consequence, the user community must participate in the project preparation and technology selection process, and the designer must know and fully understand existing conditions and user attitudes.

The planning of the sanitation component of the Jakarta Sewerage and Sanitation Project required such user participation and background information. Project authorities therefore developed a process of data collection, community consultation and statistical analysis which led to recommendations for user affordable and acceptable sanitation improvements. This process, including design of questionnaires, investigator training and computer analysis of data is described in this document in a form that will permit other project planners to utilize the process. In addition, information is provided that will enable planners to estimate the time and cost of a sanitation survey.

ABSTRAIT

Pour fournir à tous les groupes de population urbaine - riches et pauvres - des services d'alimentation en eau et d'assainissement à un prix raisonnable, il faut avoir recours à des technologies diverses, appuyées par des activités d'information et d'éducation. L'expérience a montré que les bénéficiaires, en particulier ceux qui habitent dans des quartiers sans aucun des services municipaux habituels ou sans la plupart d'entre eux, ont besoin de comprendre le coût, le but et le fonctionnement des améliorations envisagées pour pouvoir jouir des avantages sanitaires et économiques qui doivent en découler. En conséquence, la communauté doit participer à la préparation du projet et au choix de la technologie, et le concepteur doit bien connaître et comprendre les conditions en vigueur et l'attitude des utilisateurs.

La planification de l'élément assainissement du Projet d'égouts et d'assainissement de Djakarta exigeait cette participation des utilisateurs et cette connaissance des données de base. Les responsables du projet ont donc mis en place un système de collecte de données, de consultation de la communauté et d'analyse statistique qui a abouti à des recommandations sur un programme d'aménagements sanitaires acceptable et abordable pour les utilisateurs. Le présent document décrit ce processus, y compris l'élaboration du questionnaire, la formation des enquêteurs et l'analyse informatique des données, sous une forme permettant à d'autres planificateurs de projets d'utiliser ce système. Il donne également des renseignements qui permettront aux planificateurs d'évaluer la durée et le coût d'une enquête sanitaire.

EXTRACTO

El suministro de servicios de abastecimiento de agua y de saneamiento al alcance de todos los grupos de población, tanto ricos como pobres, en las zonas urbanas exige el uso de diversos recursos tecnológicos, en forma paralela con actividades de información y divulgación. La experiencia señala que los beneficiarios, especialmente los que viven en zonas con pocos o ningunos de los servicios municipales habituales, necesitan tener una buena comprensión de los objetivos, el costo y el funcionamiento de los adelantos propuestos si es que han de aprovechar las ventajas proyectadas para la salud y para la economía. Por lo tanto, los usuarios deben participar en la preparación del proyecto y en el proceso de selección tecnológica, y los planificadores deben conocer y entender cabalmente las circunstancias prevalecientes y la actitud de los usuarios.

La planificación del componente de saneamiento del proyecto de alcantarillado y saneamiento de Yakarta previó la participación de los usuarios y la información básica ya mencionadas. En consecuencia, las autoridades del proyecto elaboraron un proceso de recopilación de datos, consultas en el ámbito comunitario y análisis estadístico, del que surgieron recomendaciones para llevar a cabo mejoras en los servicios de saneamiento aceptables para los usuarios y al alcance de todos. En este documento se describe dicho proceso, incluidos la elaboración de los cuestionarios, el análisis computadorizado de los datos y la capacitación de investigadores, en una forma que permitirá su uso por otros planificadores de proyectos. Además, se proporciona información que servirá a los planificadores para calcular el tiempo y costo de un estudio de saneamiento.

TABLE OF CONTENTS

ANNEXES

DRAWINGS

PREFACE

The majority of people in developing countries do not enjoy the benefits of an adequate supply of safe water and facilities for the sanitary disposal of their wastes. Progress in improving this situation has been slow. On the one hand, funds needed for more rapid progress have not been available, on the other, conventional solutions commonly used in industrialized countries are either unaffordable to the low income groups -- the vast majority of unserved -- in developing countries or beyond their ability to operate and maintain.

This report addresses one of the most vexing problems in planning sanitation improvements in urban slum (or "poor people") areas: How to quantify the situation in a given urban area so that planning can be forcused on the most meaningful needs. What is the existing status of sanitation in the proposed project area, what facilities are already there, which are productive (or non-productive) and why, what are the gaps which must be filled in order to achieve a minimum desired level of community sanitation in the area, and how can the data be obtained so that the planner can proceed to identify and design the specific facilities needed to achieve the desired improvement? What are the users perceptions of their sanitation needs, how are they willing to participate in the improvement process?

One approach for solving this problem is based on conducting a survey of the project area to obtain the needed data. This includes not only information on the physical facilities involved but also socio-economic data relating to their acceptance and use, and it includes all of the various types of facilities which make up the overall complex of sanitation facilities in an urban slum area. The present report describes a special methodology which has been developed on how to plan and conduct such a survey, how to analyze the results, and how to utilize them for describing and specifying the needed new facilities. The methodology is based on its actual application and utilization for planning a major urban sanitation improvement project in Jakarta, Indonesia.

The successful preparation of this study, financed by the Government of Indonesia from proceeds of a World Bank loan, was greatly facilitated by the generous support the study team received from responsible officials of all Government institutions involved. The authors wish to acknowledge particularly the help of Ir. Darundono Project Manager DKI/KIP, his assistant Ir. Kapitan and responsible officials of local authorities, since the success of the survey depended to a great extent on their active support and valuable assistance. Our acknowledgements also go to Mr. Paul Blaser, Sanitary Engineer of Motor-Columbus, Swiss consulting company, who dealt with particular technical aspects of the project, as well as to all the surveyors who worked with an enthusiasm and zeal which significantly contributed to the early execution of this crucial part of the study.

We further wish to thank Messrs. L. Sud and R. Prevost, World Bank staff associated with the Jakarta Sewerage, without whose help and support the sanitation study would not have achieved its success. We are also grateful to the Banks Senior Adviser for water and wastes, J.M. Kalbermatten, who encouraged us to write this manual and who assisted in its preparation.

The authors hope that this presentation of their experience will be useful to other professionals charged with the task of improving sanitation services and will thus contribute to the achievement of the objectives of the International Water Supply and Sanitation scale.

Principal Author: Vincent Zajac, Motor-Columbus
Co-authors: Susanto Mertodiningrat, Ministry of Public Works, Jakarta
                    H. Soewasti Susanto, Ministry of Health, Jakarta
                    Harvey F. Ludwig, consultant

CHAPTER 1

INTRODUCTION

1.1 THE SANITATION PROBLEM IN URBAN LOW INCOME AREAS

One of the most difficult problems of infrastructure in the developing countries today is that of sanitation in capitals and other major urban centers where the poorer people live - usually called slum areas. Often these slum areas comprise a sizeable portion of the city's area and population, as much as half in many cases; thus they pose a formidable socio-economic problem. One of the most serious of these problems is that of sanitation, that is, the level of community cleanliness in the slum area, which in itself is a reliable indicator of the status of public health and of the quality-of-life.

The problem stems from the fact that the economic status of most residents in the slum area is not sufficient to finance conventional urban solutions to sanitation, such as piped house connections for water supply and piped sewerage. Hence dependence must be placed on use of less expensive systems which are affordable and yet furnish reasonably good solutions, such as use of individual water supply wells and of leaching pits for excreta disposal. However, because the project areas are sizeable, often containing hundreds of thousands of population, this means a multiplicity of individual systems of varying types and styles suited to a range of family income levels. Thus the overall system, instead of one single public water supply system and a single public sewer system, which are relatively easy to plan, construct, and manage comprises a vast multiplicity of individual systems which poses a formidable problem in planning, design, construction, and management.

In addition to water supply and excreta management, sanitation comprises other facilities or systems including facilities for bathing and washing, solid waste collection and disposal, surface drains, and sometimes provisions of pathways and roads so that access to homes is feasible in all types of weather. These systems again will be diverse in nature throughout the area—a multiplicity of facilities—and moreover their management often involves cooperative efforts by the residents at the local level (rather than dependence on the municipal authority), which poses another array of problems.

Despite all the complexities noted above, experience over the past several decades has clearly shown that urban slum sanitation improvement projects can be very effective in improving the environment of slum areas, because they can improve the standard-of-living in these areas at affordable costs. The planning challenge is how to handle the problem, how to quantify it so that the engineer-planner can proceed to design the needed improvements at minimum costs.

1.2 PLANNING BASED ON SURVEYS OF EXISTING CONDITION

For effective sanitation planning for the proposed project area, adequate data and information must be available on the existing facilities and their adequacy and effectiveness, and on the physical and environmental living conditions, including the health profile and the hygienic practices and preferences of the population concerned. The required data must cover the institutional, economic, and financial as well as technical aspects, and pay attention to all phases of project implementation including planning, design, construction, operation and maintenance, and monitoring. However, the collection of such data, especially when some hundreds of thousands of inhabitants are concerned, is not easy and would seem to require a great deal of time and effort. Budget constraints usually set limits to this task, hence reliance must be placed on an approach which is affordable.

A new approach to this problem was applied in a sanitation study carried out in 1982 for a portion of Jakarta, the capital of Indonesia, with an area of 7.8 sq. km and a population of about 500,000.[1] It covered all private as well as public sanitation components such as water supply (wells, piped systems), excreta disposal (toilets, leaching pits, septic tanks), bathing and washing facilities, surface drains, solid waste collection and disposal facilities and others. During a comprehensive survey of selected sample zones in the project area, carried out with the assistance of 16 local sanitary engineers, some 1,826 households were interviewed and information obtained on 963 leaching pits, 600 septic tanks, more than 150 drains, 100 public sanitation facilities, and other facilities in a survey limited to a period of one month. Altogether about 2,100 questionnaires were filled-in by the field survey crew. This produced in a short time a reliable data base on the existing problems and needs in the sanitation sector, which through analysis permitted the planning of appropriate improvements and priorities. Through use of a computer-supported system of data collection and evaluation, it was possible to minimize the time required for interviews, data processing, and extrapolation of survey results. Consequently, significant benefits in terms of time and money were achieved.

A detailed description of this approach is presented in the following chapters. This includes guidelines on (i) how to plan the survey, including design of questionnaires, training of staff, and selection of the sample survey areas, (ii) how to organize and implement the survey, (iii) how to analyze the results, (iv) how to utilize the results for planning and designing the needed improvements throughout the project area, and (v) how to estimate the time and budget requirements for a survey of this type including quantification of the amount of specialized skills which will be required.

1.3   WHY A SPECIAL SURVEY?

An important lesson learned from the experience described above is the need to explain clearly, when formulating and proposing the survey, why this particular type of survey is essential. It may be that a number of surveys have already been made of the status of the facilities in urban slum areas,

---

1.   The project, called the Jakarta Sewerage and Sanitation Project, aims to achieve comprehensive clean up of a selected pilot area or portion of Jakarta, namely the Setiabudi and Tebet kecamatans (administrative units) with a population of about 500,000. The project facilities include (i) a system of sanitary sewers serving parts of the pilot area, (ii) an improved system of individual household excreta disposal units, including improved desludging service, for the non-sewered areas, and (iii) other sanitation improvements throughout the pilot area. By demonstrating the feasibility of comprehensive clean up in the Setiabudi/Tebet pilot area, it is expected similar projects will be undertaken to cover the entire city and that the same approach can be utilized elsewhere in Indonesia.

including sanitation facilities, especially surveys carried out from the socio-economic point of view. Hence the "Decision Makers" may tend to view the proposal as another survey of the type already available. It is important to review carefully all earlier surveys, and to make use of the data they contain, and to present a solid justification for the proposed new survey. In this case, the proposed survey was undertaken to obtain information suitable for analysis to yield results which an engineer could use as the basis for delineating the specific sanitation improvements which are desired, and for proceeding to prepare the detailed plans and specifications needed for implementation.

1.4 PURPOSE OF MANUAL

The purpose of the present manual is to present, in a form convenient for use by Developing Country officials and others concerned, a practicable methodology for identifying, delineating, and quantifying the sanitation improvements needed in an urban slum area in order to achieve a desired minimum level of community sanitation and cleanliness. The methodology is based on making a limited survey of selected sample zones in the overall project area, using a trained field crew and appropriate questionnaires, to obtain the minimum needed amount of data on the existing sanitation situation. This data is processed to yield results which can be extrapolated throughout the project area, so that the specific needed improvements can be delineated together with basic design criteria, estimates of costs for construction and for operation and maintenance, and important requirements for administration and management following completion of construction. Thus the results of the survey enable the officials concerned to proceed with detailed design and implementation of the needed sanitation improvement facilities.

Through use of computerized procedures and of standardized questionnaires, the total time and cost requirements are reduced to practicle, affordable levels.

An important part of the presentation are guidelines by which officials can prepare a proposal for utilizing the survey methodology for a proposed project for a designated urban area. This includes estimates of requirements for the professional skills and other manpower and for equipment and supplies, and of the estimated survey cost. Guidelines are also given for formulating a program for continuing periodic monitoring of the sanitation situation in an urban area, to obtain information on needed improvements in basic assumptions utilized in formulating improvement projects.

1.5 PARAMETERS UTILIZED

The sanitation environment in an urban slum area is influenced by a variety of interrelated parameters which vary from city to city and from place to place within a city. These include not only the physical facilities per se but the acceptance and use of these facilities by the intended beneficiaries and by others, and also the extent to which the beneficiaries are willing to contribute to installing and maintaining the facilities either through payments or contribution of services.

For purposes of the present study the following are considered to be the salient parameters:

(a) Existing Physical Facilities

Types, sizes, capacities, amounts, locations, sketches (as needed to quantify the facility from the engineering point of view):

(i) Water Supply (wells, public taps, connections to piped systems) (amounts produced/used) (safety and quality of supply).

(ii) Excreta Management (toilet and plumbing facilities) (individual disposal units of various types, connections to public sewers or drains).

(iii) Communal Units for Washing/Bathing/Toilets.

(iv) Solid Waste Management (collection, storage, transfer, and disposal).

(v) Surface Drainage (surface drains, special drainage problems such as areas of low elevation).

(vi) Others (including access ways and related desludging problems).

(b) Operation and Maintenance

(i) Acceptance and Use (extent to which the facilities are being or are not being used as planned and why) (extent of use by non-intended beneficiaries).

(ii) State of Repair of Facilities (condition of facilities, needs for repair, records and evidences of repairs).

(iii) Administration (who has responsibility) (adequacy of O&M with explanations) (role of householder and of local political chiefs and other officials) (role of vendors or other middle-men) (contributions by beneficiaries in money or services) (deficiencies) (extent of fees and of compensation to maintenance personnel and who sets them).

(iv) Levels of Service (per capita use of water, both total and for various purposes) (adequacy of toilet and excreta disposal facilities for meeting family needs) (adequacy of solid waste management service, adequacy of drainage, etc.) (for various categories of population by income levels, including squatters).

(c) Institutional Support

What other institutions contribute to the success of planned improvement? (Health agencies, clinics in teaching value of personal hygiene; local clinics monitoring health status; health officers checking water quality; schools teaching health/water/sanitation/nutrition relationships; adult education teachers and community workers organizing and motivating users in the construction and maintenance of facilities)? How are the actions of these and the project agency planned and coordinated?

(d)   Impacts on Environment

Adequacy of facilities for protecting public health and other environmental values; water contamination and pollution hazards; insect and rodent vector hazards; environmental cleanliness.

(e)   Periodic Monitoring

Whether any agencies make periodic evaluations of the performance, status, adequacy, and acceptance and use of the facilities, and if so, what reports are produced and who gets them.

While the listing above serves to illustrate the complexity of the problem, through the use of sample areas, standarized questionnaires, and computerized processing it is possible to obtain a quantified description of the overall existing situation and of the extent to which it is meeting the needs.

1.6   SCOPE OF MANUAL

In summary, the present manual aims to do the following:
(a)   Describe a computerized methodology (i) for conducting a limited survey of sanitation parameters in an urban slum area proposed for sanitation improvements, (ii) for analyzing the data to yield results which can be extrapolated throughout the project area, and (iii) for using the extrapolated findings as the basis for designing and costing the specific sanitation improvements needed for achieving the desired minimum level of community sanitation, including provisions for operation and maintenance, and including institutional as well as technical aspects involved in administration and management.

(b)   Illustrate the use of the methodology by describing its application to an urban slum area in Jakarta.

(c)   Present guidelines for formulating a survey of this type for assisting in the planning of proposed new sanitation projects in urban slum areas, including costing.

There is a need also for formulating guidelines on how to establish and maintain an optimal program for continuing periodic monitoring of sanitation system performance in urban slum areas, to obtain feedback information needed for improvements in planning/design criteria as well as for needs for physical repairs and for improvements in provisions for administration and management. This task is not included in the present manual because the periodic sanitation monitoring program for the Jakarta Sewerage and Sanitation Project is yet to be implemented. This will be done within the completion of construction of the needed sanitation improvements.

CHAPTER 2

SURVEY PREPARATION

2.1   BACKGROUND DATA REQUIREMENTS

Before starting any field work, all available relevant data and documentation related to the area to be surveyed have to be collected, classified and analyzed, in order to obtain the best possible background information base. This is necessary for preparing the survey, setting up its objectives, delineating the parameters, selecting the sample survey zones, and focusing the investigations on the main problems. Particularly required for this purpose are data and documents concerning:

(a)   The topographical, geographical, geological, political, demographical, and socio-cultural features of the study area.

(b)   The history of economic and social development of the area as well as planning for future development programs.

(c)   Current overall living standards and the primary sanitation, health, and environmental problems within the area.

In this connection, the following documents are essential:

(a)   Administrative maps (as detailed as possible) delineating political boundaries within the area and the administrative structure showing the responsibilities of appropriate national, regional, and local authorities.

(b)   Geographical and topographical maps with a scale of at least 1:10,000, but preferably 1:5,000 or less, indicating rivers, streams, lakes, swamps, ponds, roads, streets, railways, canals, bridges and other relevant data with appropriate topographical contours within the area.

(c)   Aerial photographs with a scale of at least 1:10,000, where the settlement density, roads, and housing patterns may be examined (using a steroscope). With aerial photographs of 1:5,000, these features can be easily distinguished by eye or with a simple magnifying glass).

(d)   Hydrogeological maps (as detailed as possible) delineating the ground water level in meters above sea level and - for coastal areas - ground water isochloride curves for wells within the area.

(e)   Geological maps (as detailed as possible) showing soil qualities, particularly the soil permeability, within the area.

(f)   Demographic statistics (census data) for basic administrative units, comprising total population and its distribution, growth rates and migration movements, birth and mortality ratios, age and sex structure, religious, ethnological and social patterns, economic activities, location of squatters within the area, and government policies related to management of squatters.

(g)   Health statistics, including morbidity and mortality rates, especially gastro-enteric diseases and other water-borne diseases, existing health services and facilities, medical staff, and preventive health services.

(h)' Plans of existing and proposed piped water supply systems (including information on sources of supply) distribution networks and mains, deep-wells, public water taps, and hydrants together with information on institutional, managerial, and operation and maintenance aspects, capital and operation costs, water production, consumption, and prices.

(i) Information (including standards and locations) of existing and proposed communal sanitation facilities (toilets, bathing, washing, etc.), their managerial, operation and maintenance systems, capital and operation costs.

(j) Practices and standards of existing house sanitation facilities (water supply, toilet system, sullage and solid waste disposal, etc.), their capital cost and maintenance requirements.

(k) Information on existing desludging equipment, system and routes of desludging services applicable to the study area.

(l) Standards and status of existing and proposed roads, footpaths, sewers and surface drains (canals and ditches), their maintenance systems, capital and operation costs.

(m) Information on existing and proposed important public and/or private institutions (schools, mosques and churches, industries, hotels, offices, markets, stores, farms, etc.).

(n) Other development programs (short, medium, and long-term) within the area, particularly in the sectors of housing, transport (railways, highways, roads), and power.

(o) Previous studies, reports and other relevant publications concerning sanitation aspects in the study area, including health and overall environmental problems, sanitation improvement programs implemented or under implementation, etc.

The sources and the availability of the background data vary from one country to another and from city to city. For larger cities most of the data required are normally available, since they are essentially basic prerequisites for town planning activities anywhere. In Jakarta, for example, all the above documents were available; mostly, they were supplied by the Municipality and other government agencies. The fieldwork in the project area was considerably facilitated by the fact that the city maps and aerial photographs of scale 1:5,000, plus some maps even with scales of 1:2,000 and 1:1,000 (for selected sample survey zones), were supplied for the study team.

Where the above data and documentation is not available, or only in parts, the main sources of information would be appropriate local leaders and officials within the project area, particularly those involved in community development planning, health and social affairs, housing, water supply, electrification, education, road construction, and transportation. Important information could be obtained also from persons who are dealing, directly or indirectly, with the people concerned, with their problems, or with any other aspects relevant to the objectives of the study; such persons could be for example, teachers, physicians, political and religious leaders, social workers, local contractors, transporters and other professionals leaving or executing their activities within the study area. For the field work, very probably more sketches than maps (or excerpts of maps) will be used. For these purposes more preliminary field investigations will be necessary early in the preparatory phase, especially for classifying the area, selecting and delineating sample survey zones, identifying the topographical particularities (e.g. flooded zones) of the area, specifying the main problems on which the survey should be focused, as well as for carrying the actual field work. The preparatory work in such a case will require more time and, consequently, appropriate provisions should be allocated in the study budget for this purpose.

2.2 SELECTION OF SAMPLE SURVEY ZONES

The area to be surveyed, especially in large urban centers, is usually very heterogeneous with regard to housing patterns, density of population, family income levels, road accessibility, area topography and other settlement aspects. In such a case, to have a firm basis for extrapolation of sample survey results to the whole project area, particular attention should be paid to the selection of sample survey zones, in order to achieve the highest possible degree of reliability and representativeness for the data collected.

For this purpose, it is recommended first to divide the project area into sub-areas and to classify them according to prevailing housing patterns (corresponding practically to different family income groups), population density, distance to rivers, flood risks, road accessibility and other relevant criteria. As is well known, this settlement diversity influences the hygienic behavior of the population and various health and overall environmental problems.

By dividing the project area into sub-areas, it is helpful, for practical reasons, to follow political boundaries to the maximum possible extent. Identical boundaries of a survey sub-area and an administrative unit facilitates the fieldwork of surveyors (support of local authorities), the extrapolation of survey results (generally, population data are available by administrative unit only), and the planning of improvement measures, which is usually required for each administrative unit involved. The main principle to be followed in this work is, of course, high settlement homogeneity within the sub-area, no matter what its size.

For the classification of sub-areas, the above mentioned geographical, topographical and geological maps, the aerial photographs and plans of piped water supply systems can be used. In addition, a preliminary investigation of the project area focusing on housing patterns, topographical particularities, population density and road accessibility, appears very useful if not indispensable. Valuable information (e.g., identification of flood zone areas, main sanitation problems, etc.) can be obtained by preliminary interviews with well-informed officials of local authorities within the area.

Thereafter, once all sub-areas are appropriately classified, representative sample survey zones for the actual field investigation are selected and delimited, whereby for each type of classified sub-area, one or more sample survey zones shall be considered, depending on the size of the area, number of sub-areas and their classification types, and on available time and funds.

In the case of the Jakarta Sewerage and Sanitation Project, aiming at the general improvement of the environmental health conditions of all inhabitants within a large portion of the city, particularly those living in "poor" areas, called "kampungs", the project area was divided into 56 sub-areas, of which 43 were kampungs and 13 were non-kampung sub-areas.[1]/ Administrative boundaries were fully respected (i.e. sub-areas did not cross administrative boundaries).

For classification of these sub-areas, the following types of prevailing housing standards were used:

(a) Temporary and transient housing (bamboo-wood structure).

(b) Semi-permanent housing (solid foundation, temporary structure).

(c) Permanent housing (solid structure).

(d) Individual modest housing.

(e) Individual high-income housing.

(f) Residential area with small-scale industrial and/or commercial undertaking.

Each type was further classified as:

(a) High-ground area.

(b) Flood-prone area.

Together, 24 sample survey zones were selected covering all types of classified areas. Their delineation was fixed during the preliminary investigation of the area, and at the same time some of the data needed for the practical organization of the actual field survey (road accessibility, location of local authorities involved, sites for parking of survey team's vehicles, etc.) were checked in situ.

For illustration, maps showing the study area with political boundaries, different types of sub-areas and selected sample survey zones are presented in Drawings 1 to 3, respectively.

2.3 PREPARATION AND TESTING OF QUESTIONNAIRES

The questionnaires to be used for interviews or observations in the field need to be prepared for different sanitation components separately, i.e. for:

1/ The size of sub-areas varied from 2 to 115 ha, the number of inhabitants from 500 to about 26,000, with a density of population from 200/ha to 937/ha.

(a) House sanitation facilities, including water supply, toilet systems, septic tanks, leaching/soakage pits, sullage and solid waste disposal, and hygienic behavior of family members, when all relevant data and information related to the household can also be collected, such as:

(i) Housing and family patterns.

(ii) Family health status.

(iii) Family living standards.

(iv) Family attitudes towards public sanitation facilities.

(b) Public water taps relying on municipality or local piped distribution systems.

(c) Communal sanitation facilities, including toilet, washing, bathing and laundry units.

(d) Surface drains, including canals and ditches.

(e) Communal solid waste disposal facilities.

This gives a flexibility for organizing the field work (division of tasks among different members of the survey team—see section 3.2) and facilities evaluation of data collected (see Chapter 4).

Such questionnaires, as used in the Jakarta study, which includes all the above sanitation components, except communal solid waste disposal[1], are presented in Annex 1 (house sanitation facilities), Annex 2 (public water taps), Annex 3 (communal sanitation facilities), an Annex 4 (surface drains). These questionnaires were first translated into the local (Bahasa Indonesia) language, before their actual use by surveyors in the field. They include some improvements resulting from the survey experience. For illustration of the actual work, selected copies of completed questionnaires, including sketches of surveyed facilities, are included.

The questionnaires with coded answers are designed in such a way that the data obtained can be analyzed by a computer program, hence the time required for interviews and observations as well as for input data preparation are reduced significantly. Surveyors do not need to write answers; they only record appropriate answer codes. The completed questionnaires are then used directly as input data sheets for processing.

The questionnaires of course have to be translated into the local language and appropriately adjusted to the conditions of the new project area as well as to the objectives of the study. Further, before their actual use in the field, the prepared questionnaires must be tested, as was the case in the Jakarta study, by interviews with a number of families and local officials, in order to check that the questions and anticipated answers are easily comprehensible to the persons to be interviewed, correspond to local condi-

---

1/ Community solid waste disposal was dealt with separately as part of another project.

tions, and cover fully all aspects of existing sanitation facilities in the area. Only after that, the tested and, if necessary, adjusted questionnaires should be reproduced in sufficient quantities for the survey team.

2.4 SETTING UP AND TRAINING OF SURVEY TEAM

The selection of surveyors is crucial for the success of the survey, since the working conditions in large, overcrowded urban slum areas are generally difficult, especially if there are serious hygienic and environmental problems. The study team should consist of specialists with qualities of motivation and dedication, who are familiar with the living conditions of the people in the area.

Experience has shown that junior sanitary engineers or university students of this category can be most efficient for such survey work. When the surveyors are socially and culturally similar to the population with whom they will work, communication is even more effective. In the case of the Jakarta study, all surveyors were young sanitary engineers, were well acquainted with life in kampungs (some of them lived there), and had a personal interest in gaining practical experience. This contributed significantly to a correct respect for the survey's approach and schedule, in spite of difficult condition in the field.

Once the survey team is set up, training should be organized to explain the aims, the concept and methodology of the survey, i.e., how to start the work, how to interview people, how to make observations and, finally, how to fill in the questionnaires. Thereafter, practical training must be undertaken in the field for each sanitation component: house sanitation facilities, public water taps, communal sanitation facilities, surface drains, etc. This helps, at the same time, to check if the surveyors themselves have understood the methods and techniques to be applied and, thus, to see if they can work independently. An essential aspect of the training program is to have different teams survey the same area at different times, to be sure the results are consistent and reproducible. The reason for doing this should be carefully explained to the households to be interviewed and their prior approval for a second visit obtained.

Efforts have to be made at the beginning of the contact with the household to establish a climate of understanding and cooperation between surveyors and interviewed persons. A general informal discussion with family members about their activities, health, problems, etc., before any formal interview following the prepared questionnaires could help to create a favorable relationship between surveyors and interviewed persons and consequently to achieve the desired climate of an open and frank cooperation.

In Jakarta, the survey team consisted of 10 sanitary engineers from the Municipality staff and 6 university students from the Sanitary Engineering Faculty in Bandung. They were divided into 8 working groups, each consisting of two surveyors, with one appointed as group leader. Their work in the field was coordinated by Ir. J. Wiriadipura, sanitary engineer of Encona, a local consulting firm, assigned as survey team leader, under the supervision of and in close cooperation with Ir. Kapitan from the Jakarta Municipality. The preparation, organization and execution of the work were managed by Dr. V. Zajac, senior economist-engineer of Motor-Columbus, the Swiss consulting company which furnished outside expertise, with assistance from Dr. Harvey

F. Ludwig, consultant for the World Bank.

The training of the surveyors lasted 3 days. During the classroom training, all four types of questionnaire were discussed, question by question, and appropriate explanations were given. Explanatory notes, drawn from this experience, are attached to the questionnaires presented in Annexes 1,2,3 and 4.

For the training in the field, the survey team was divided into four training units, each consisting of four surveyors (2 working groups), in order to avoid overcrowding during interviews with households and blocking traffic during the work in the streets. Each training unit, supervised and assisted by the above-mentioned specialists, surveyed successively all four sanitation components (house sanitation facilities, public water taps, communal sanitation facilities, and surface drains), according to the prepared training program (See Annex 5).

CHAPTER 3

ORGANIZATION AND EXECUTION

3.1 ORGANIZATION AND MANAGEMENT OF SURVEY

The institutional, managerial and operational responsibilities for the survey must be delineated and a survey organization chart set up to guide the survey operations. This should include a clear designation of the responsibilities of and relationships between different institutions and agencies involved in the study as well as different categories of staff in charge of the execution of the project. Because the sanitation problems within a city are usually dealt with by the Municipality, it can be assumed that any sanitation survey of urban slum areas will be under the responsibility of the Municipality, even if the project itself is to be implemented by some other government agency. Once the managerial staff for the sanitation study has been assigned, a schedule for the whole survey period, including preparatory work, training of the survey team, survey execution, and processing and evaluation of data collected must be established, according to the objectives of the study and available funds.

Further, appropriate documents, working material and equipment necessary for the actual survey must be prepared. These include:

(a) Maps or sketches of sub-areas with selected survey zones and delineation of streets for interviews with families.

(b) Maps or sketches of sub-areas with location of public sanitation facilities (communal sanitary blocks, public water taps, public solid waste disposal facilities) and surface drains to be surveyed.

(c) Water pressure gauges for measuring the pressure of water in water taps.

(d) Acoustic and/or light level gauges for measuring the water level in wells.

(e) Metallic tapes for measuring surface drains, dug wells, and other purposes.

(f) Auxiliary field materials such as desk pads for questionnaires, maps and sketches, pencils and erasers.

In addition, suitable rooms for meetings and office work with secretarial and drafting services, as well as an adequate number of vehicles for fieldwork have to be provided. Since in urban slum areas the capacity of motorable roads is generally very limited and traffic is often congested, it is preferable to use a number of small-capacity vehicles rather than one big one for the transportation of the survey team. This provides greater mobility and flexibility for organizing the work.

In the case of the Jakarta study, as illustrated by its organization chart presented in Annex 6, the sanitation survey was carried out under the responsibility and supervision of the Municipality, although the implementation of the project as a whole (due to the sizeable scale of sewerage components in terms of total budget) remained under the responsibility of the Ministry of Public Works. This direct involvement of the Municipality was required especially because a similar sanitation improvement program, called the "Kampung Improvement Program" (KIP), involving financing by IBRD, had already been underway for many years under a special municipal department.

The schedule of the Jakarta Sanitation Survey is shown in Annex 7. As can be seen, the overall sanitation study required three months but the actual field survey about one month only. The fieldwork was organized in such a way that one sample survey zone, including the public sanitation facilities within this zone, and adjacent sub-areas were investigated in one day. For each daily trip, a survey program with assigned tasks for each working group was set up and distributed to the survey team before its departure to the field. For this purpose, valuable documents (street maps and drawings of existing surface drains, standards and location of various public sanitation facilities) were supplied by the Municipality's KIP unit. For illustration, a selection of such maps is presented in Drawings 4 to 6.

For transportation of the survey team, two minibuses (capacity of 10 persons each) and one car (for 4 persons) were rented, with drivers. A meeting room with a blackboard for classroom training and daily meetings with the surveyors was made available by the Municipality. The team was equipped with eight water pressure gauges, one acoustic and one light level gauge, and each surveyor with a metallic tape measure, desk pad, pencil and eraser.

3.2 EXECUTION OF SURVEY

In principle, two methods could be used for the execution of the survey:

(a) To divide the survey team into three or four groups, whereby one or two of them would survey all house sanitation facilities, and another one public water taps, and communal sanitation facilities, and the last one surface drains, or

(b) To charge the team to survey, successively, all sanitation components considered.

Because the working conditions for surveyors may vary from one sanitation component to another, it is preferable to use the latter method. This permits each surveyor, although to a different extent, to deal with all sanitation components. In addition, this method facilitates the transportation of the survey team to, within, and from the area.

In any case, the survey team should work in 2-man working groups. This considerably facilitates the work (while one surveyor is interviewing or measuring, another is filling in the questionnaire) and helps ensure that the survey's approach is respected and that the expected reliability of data collection is achieved.

It is very important that the survey team always be accompanied to the field by the supervisory staff (manager and/or supervisor of the survey) to facilitate contact with local authorities, to obtain their support, to delineate tasks in situ for each working group, and to ensure smooth progress of the work. Further, it appears advantageous to hold a short meeting with the survey team everyday to collect the filled-in questionnaires, to discuss the experience from the fieldwork, and to distribute the questionnaires with related documents for the next trip. This also allows the checking of the progress of the work and the settling without delay of any problems which might appear.

During the survey itself, it is appropriate, if the families interviewed are not in a position to answer some questions, that qualified surveyors make their own estimates on the basis of their own observation and knowledge, for example, of family incomes, quantity of water consumed, etc. It also helps very appreciably, if the surveyors of drains, public sanitary facilities and water taps record their own recommendations for improvements; this facilitates the analysis of survey results and preparation of proposals.

In Jakarta, the second method (b) was applied. Besides the above-mentioned factors, including equal working conditions for each surveyor and easy transportation, it was the strong motivation of young sanitary engineers to learn in practice about all types of sanitation facilities which favored this method.

Since the surveyors lived in different parts of the city, they met every morning at 8:00 am in the Municipality meeting room, from where they were transported to the survey area. Before their departure to the field, the filled-in questionnaires from the previous day were collected and new ones for the following trip distributed. The experience and problems encountered were discussed and explanations were given as required. After arriving in the area, at the preselected parking place, the manager and supervisor visited the responsible officials of the kelurahans[1]/, who had been informed in advance by the Municipality about the survey, and asked for their support (See Section 3.3).

Thereafter, assisted usually by a local official, the study team (eight 2-man groups) started the work of interviewing households, each working group visiting houses along the streets indicated in the maps/sketches prepared by the survey manager. When the survey of the zone was finished (about 10 households per group per day), each group proceeded to the investigation of surface drains, along the preselected roads and paths, always according to the prepared street maps or sketches. As shown on Drawing 6, each group was charged to survey one or two sections of surface drains located and selected so that all sections, surveyed by eight groups, formed an integrated part of the existing drainage system within the surveyed sub-area. After this, the public water taps and communal sanitation facilities within the sample survey zones and adjacent sub-areas were investigated, each group following the tasks assigned by the survey manager in the prepared maps or sketches. In this

---

1/ Kelurahan is an administrative unit of the city, headed by a Lurah (Chairman).

connection, the managers and/or operators of the surveyed facilities were interviewed.

Once the work tasks noted above were accomplished, the survey team returned to the vehicles and were transported home, after about 5 to 6 hours of fieldwork.

To have complete information about all existing public water taps and communal sanitary facilities within the project area, the remaining facilities located outside the sample survey zones were investigated by a few of the surveyors after the sample survey of house sanitation facilities and surface drains was completed.

In addition, to obtain realistic information about the effectiveness of desludging services, a sanitary engineer from the Jakarta Sewerage and Sanitation Project staff accompanied the operators of a desludging crew during one day's work. One of the small vacuum trucks with a 2.4-ton capacity was used. Also, in order to check the efficiency of desludging services in terms of solids removal, tests were carried out in cooperation with the Municipality Cleansing Department. Desludging of four selected leaching pits was monitored by measuring the drawdown of the liquid level and analyzing the dry residual of sludge samples in the pit after stirring and before desludging, after desludging, and in the vacuum tank. In addition the depth of ground water level was measured in a nearby dug well.

When all field investigations were completed, a closing discussion with the survey team was organized in the Municipality meeting room. The experience from the work accomplished was evaluated and, at the same time, some outputs of the computerized data already available were demonstrated to the surveyors. The officials of the Municipality appreciated the work of the surveyors and, as acknowledgement of their efforts and zeal, a special remuneration was granted to them on completion of the surveys.

3.3 COOPERATION WITH LOCAL AUTHORITIES

A very important factor is the support of local authorities involved. They should always be informed in advance about the objective of the job and the advantages the project will bring to the people. The work of the surveyors is facilitated if they are provided with official letters of introduction and accompanied in the field, at least at the beginning of the work, by appropriate official staff.

At the beginning of the study, Ir. Darundono, KIP Project Manager for the Municipality, invited officials of the kecamatans involved (Setiabudi and Tebet) to a meeting in the Municipality, informed them about the objectives of the survey and asked them to support the survey team during its work within their areas. Subsequently, the chairmen of these kecamatans transmitted this information and an appeal for cooperation in a circular letter to their subordinate local officials (lurahs, RW, RT)[1] and, at the same time, they

---

[1] RT and RW are designations for the basic local administrative units, and their chiefs with the main function of transmitting information from or to the people and Government Authorities (Scheme of authority: RT - RW - lurah - Camat - Wali Kota (City Mayor) - Governor). The RT is responsible for about 50 households (neighborhoods) and the RW for 10 to 20 RTs. The RT and RW chiefs are selected by the citizens in their areas.

issued letters of introduction for the survey team. This helped considerably in obtaining advance information about the location of flooded areas, the frequency and extent of innundations, the location of existing and proposed public water taps, deep-well stations, communal sanitation facilities, number of inhabitants per basic administrative unit, road accessibility and parking possibilities within the surveyed areas, and other relevant aspects related to the study.

As mentioned in Section 3.2, during the fieldwork local official staff accompanied the study team in the survey zones and facilitated contacts with the population. Valuable information and suggestions were obtained from local officials with regard to the operation, maintenance and effective use of the communal sanitation facilities, public water taps, and surface drains surveyed, the hygienic behavior and attitudes of the population concerned, and particular sanitation problems within the area of their responsibilities.

At the end of the study, the survey's findings and conclusions as well as the proposed sanitation improvements were discussed with local officials with the following objectives:

(a) To inform them about the results of the survey .

(b) To check the acceptability of the proposed improvement measures.

(c) To obtain their support for the implementation of these measures, particularly by making land available for new facilities, by organizing self-help operation and maintenance services for communal sanitation facilities and surface drains, and by helping with an educational campaign for the people affected.

CHAPTER 4

ANALYSIS OF RESULTS

4.1 PROCESSING OF SURVEY DATA

Two computer programs are currently available for data processing:

(a) SPSS (Statistical Package for the Social Sciences) Program, prepared by SPSS Ltd., 444 North Michigan Avenue, Chicago, Ill. 60611, USA. It is conceived principally for demographic census, and similar socio-economic studies, but is easily adaptable for the purposes of a sanitation survey.

(b) SODEMOS (Social Demographic Survey Evaluation) Program, developed by Motor-Columbus Consulting Engineers Inc., Parkstr. 27, Baden, Switzerland. This program was prepared specifically for sanitation surveys, but is suitable for other socio-demographic studies, such as resettlement and transmigration studies, water supply projects, and health development projects.

The SPSS Program is generally used in developing countries for demographic census purposes and is usually available from existing computer centers in all these countries. The SODEMOS Program can be obtained from Motor-Columbus.

For data processing, it is recommended to do the punching of the input data step by step, following the progress of the survey, in order that this time-consuming work can be completed immediately after the final investigation of the survey area is finished. The completed questionnaires should be checked to see if the codes are written correctly, to avoid mistakes or misinterpretations during punching. If mistakes are found, it is best to clarify these with the appropriate surveyor during the regular daily meetings of the survey team. For this reason, particular attention must be paid to writing the codes, i.e., to record easily understandable figures (0 -9).

After checking the punched input data, the subsequent processing takes only a short time and the data can then be aggregated into packages according to the extrapolation requirements, e.g., for different types of sample survey zones and for different administrative units as well as for the entire project area.

In Jakarta, the survey data were processed in the Municipality computer center. Because this center already owned the SPSS program, this general program was used. The completed and checked questionnaires from each sample survey zone were given immediately for punching and processing, which enabled the output data for the whole survey to be available shortly after the last investigation in the area had been completed. Altogether 41 packages were processed as follows:

(a) For house sanitation facilities (total: 1,826 questionnaires):

24 individual sample survey zones.
5 sample survey zones combined as follows:
Kampungs in Kecamatan Setiabudi,
Kampungs in Kecamatan Tebet,

All kampungs ,
Non-kampung areas ,
Total project area .

(b)   For surface drains (151 questionnaires):

10 sample survey zones combined, per kecamatan, as follows:
Kampungs with low-income groups,
Kampungs with medium-income groups,
Kampungs with high-income groups,
Non-kampung areas ,
Total kecamatan's area .

(c)   For public water taps (66 questionnaires).

(d)   For communal sanitation facilities (24 questionnaires).

For illustration, some selected out-put data are presented in Annexes 8 to 12.

4.2   EXTRAPOLATION OF SURVEY RESULTS

The extrapolation of the computerized survey results for the whole project area could be based on population data, number of houses, or other factors, depending on the facilities concerned, whereby the results from particular sample survey zones should be applied only to corresponding (similar) types of project sub-areas, in order to achieve accuracy and reliability in extrapolation.

As shown in Annexes 13 and 14, the extrapolation of survey results for house sanitation facilities in the Jakarta project followed this principle (compare Drawings 2 and 3 showing different types of sub-areas and selected sample survey zones). The extrapolation for surface drains was done for four main categories of sub-area (based essentially on different income groups of the population concerned), for the purpose of evaluating the possibilities for self-help and remuneration for maintaining micro-drains or ditches within the project area.

The extrapolations for housing and family patterns was based on population data for kampung and non-kampung areas and those for water supply, toilet systems, septic tanks, leaching pits, and solid waste disposal on the number of houses. The latter was obtained by dividing the population of the area by the average number of occupants per house resulting from the survey. The extrapolation for micro-drains was based partly on the length of such drains already existing in some areas, partly on lengths obtained by multiplying the sub-area surface (in ha) by an average length of drains per ha from similar sub-areas.

For illustration of these extrapolations, some selected data related to the house sanitation facilities within the project area are presented in Annexes 15 to 17, and those concerning surface drains in Annex 18. The evaluations of data concerning existing public water taps and communal sanitation facilities (completely surveyed) are presented in Annexes 19 and 20, respectively.

4.3 QUANTIFICATION OF NEEDED IMPROVEMENTS

To specify and propose needed sanitation improvements, problems and gaps encountered have first to be identified and quantified. In this connection, the most relevant data obtained from the survey would be as follows[1]:

(a) Water Supply:

(i) How many people (and where) use water for drinking and cooking from shallow wells, dug wells and/or other unsuitable water sources?

(ii) What is the distance of these wells from the nearest pit latrine, toilet, effluent drain field or other waste treatment system? Is this distance sufficient to avoid well contamination hazards?

(iii) Is the capacity of existing public water supply systems (piped systems, deep-well stations, public water taps/ hydrants) sufficient and are their operation and main- tenance satisfactory to meet the water demand within the area/sub-area?

(iv) What would be the most appropriate water supply system for furnishing the most socially and economically acceptable level of services?

(b) Excreta Disposal Systems:

(i) How many people (and where) are without any toilet system in their house? In such cases, where do they defecate?

(ii) If a communal latrine is not used in such cases, what is the reason? Is it too far? Is the fee too high? Lack of privacy? Lack of appropriate maintenance?

(iii) If the communal latrine is too far, how far are people willing to walk?

(iv) If the fee is too high, how much are they able and willing to pay for the use of a toilet?

(v) What improvements to existing facilities and/or what new facility design can be expected to meet privacy requirements?

(vi) How can maintenance be improved to keep the facility clean? Are people willing to maintain (clean) it themselves (by organized self-help)?

----

[1] All other data collected could be, of course, used as valuable complementary supporting information for detailed planning purposes.

(vii)   What else should be done to make communal toilets acceptable?

(viii) How many houses within the area/sub-area have toilet systems from which used water goes directly to drains, ditches, rivers or open land?

(c)   Desludging Services:

(i)   How many leaching pits and/or septic tanks with leaching systems are within the area/sub-area and where?

(ii)   Are these facilities regularly emptied/desludged? If not, why? Are they accessible for desludging equipment in use? Is the cost of desludging services too high? Is the desludging capacity sufficient?

(iii) What is the maximum distance from houses to roads making desludging by equipment currently in use possible? How many houses are located beyond this distance? What new desludging technology (e.g., small vacuum trailers) could be used to meet desludging requirements within the areas/sub-areas inaccessible to desludging equipment currently in use?

(iv)   If the desludging cost is too high, how much are people able and willing to pay for it?

(v)   Is the functioning of leaching pits impaired by excessively tight soils and/or by high ground water levels? If so, quantify the situation.

(vi)   If the capacity of existing leaching pits is too small, requiring a higher desludging frequency, what design (capacity) modifications of these facilities is needed to reduce desludging frequency to an acceptable level.

(vii) Are the desludging trucks operated by the Municipality willing to give service to the poor people areas or do they tend to avoid these areas and restrict operations to higher-income level areas?

(viii) What measures need to be undertaken to make desludging services satisfactory?

(d)   Surface Drains:

(i)   What is the length of existing surface drains requiring larger capacities and repairs?

(ii)   What is the length of new drains to be constructed within the area/sub-area?

(iii) Is the maintenance (cleaning) of drains adequate? If not, why (lack of funds, use of drains for disposal of refuse by local inhabitants, or other reasons).

(iv)   If lack of funds, are people willing to maintain (clean) the drains themselves (by organized self-help)?

On the basis of such data, obtained through extrapolation of sample survey results (home sanitation facilities and surface drains) or directly from investigations (public water taps and communal sanitation facilities), improvement needs can then be easily quantified.

Thereafter, using the planning parameters and criteria usually applied for different sanitation facilities, appropriate sanitation improvements can be proposed including provision of physical facilities with related implementation schedules, capital and operation and maintenance cost estimates, as well as corresponding managerial, institutional and financial measures necessary for achieving these improvements, in accordance with government policies in this sector and the funds available for the project.

It is understood that the costs of investments proposed, as well as operation and maintenance costs, will vary from one country to another, depending not only on the technologies selected but also (for the same technology) on the differences in local cost components (land, building materials, manpower, energy, water, etc.), the transport costs of equipment and materials to be imported, inflation rates and other factors. This applies to basic cost estimates as well as physical contingencies, price contingencies, and engineering costs. For financing purposes, an appropriate breakdown of these costs into foreign and local cost components as well as the establishment of a disbursement plan is usually required.

The Jakarta study confirmed that large-scale efforts have already been made and are still underway by the government to improve health and general living conditions of the population in large city slum areas. Through the construction of thousands of kilometers of canals and ditches along the streets and footpaths, hundreds of public water taps relying on the piped water distribution systems or deep wells, thousands of individual leaching pits, hundreds of public sanitation facilities (including public washing/bathing/toilet facilities called "MCKs"), and solid waste management facilities, a significant level of overall environmental cleanliness in kampungs has been attained. Nevertheless, in spite of these very considerable achievements, the following main sanitation problems were still encountered within the project area during the survey in Setiabudi and Tebet Kecamatans (See tabulations in Annexes 16 to 20):

(a) Inadequacy of safe water supply, particularly for poor people:

   (i) About 176,000 persons, over 56% of the population, use drinking water from shallow wells and dug wells (with high risk of contamination).

   (ii) Over 75% of public water taps (about 100) have no water. In the remaining ones (27), the pressure is very low.

(b) Lack of public sanitation facilities:

   (i) There are still over 35,000 people without an in-house toilet system; about 8,000 of them use MCKs, the remainder use neighbor's toilets, or rivers, ditches, drains and open land.

(ii) Existing MCKs have many deficiencies and require upgrading as well as improvement of operation and maintenance.

(c) Lack of appropriate individual excreta disposal:

(i) There are about 3,800 houses with toilets without any leaching system and these toilets overflow to drains, ditches, rivers or open land. Even after the proposed construction of some sewers in the area, toilet discharges from about 3,000 houses would still contaminate the environment.

(ii) Many of the 32,000 existing leaching pits are too small and are not or cannot be desludged due to financial or technical problems such as inability of poor families to pay desludging costs and inaccessibility for existing desludging vacuum trucks equipped with an 80 m hose. Detailed analysis of accessibility (see Drawing 12) has shown that even with longer hoses (e.g., 120 m) all areas could still not be reached.

(d) Poor quality of surface micro-drains:

(i) Some 177 km (33.7%) of these drains require repair and 24 km of new drains have to be constructed.

(ii) Drains (inadequately maintained) serve not only for removing storm water, but also sullage water and discharges from toilets (6.6% of the total number of houses), as well as solid wastes, all of which produce health hazards for the population.

CHAPTER 5

DESIGN OF IMPROVEMENTS

5.1 PRELIMINARY DESIGN OF IMPROVEMENT MEASURES

To remedy this situation and to achieve, together with the introduction of a new sewerage system into the area, a further betterment of the environmental health and general living conditions of the people concerned, particularly in poor kampungs, the following sanitation improvements within the next three-year period were proposed:

(a) Water Supply

(i) To rehabilitate/reactivate all 100 public water taps now out of operation.

(ii) To establish an additional 100 water taps with appropriate extensions (12,000 m) of the piped distribution system relying on deep wells (8). (See tabulation in Annex 21 and Drawing 7)

Assuming rehabilitation of the 100 taps out of operation, the need for new water taps (to meet the demand in the next ten-year period) was calculated on the basis of 1,000 persons per tap with 2 faucets, i.e., about 500 persons/faucet (even though 200 persons per tap was used for preliminary design of the recommended improvements). In the calculation, the following factors were taken into account:

(i) The average number of persons currently using water from existing taps in operation is 714 persons (See Annex 19).

(ii) With an increase in living standards, more families will shift to house connections and people still dependent on the use of public taps will be spread throughout the kampungs.

For the extension of the distribution network, the same criteria as used in similar previous projects were applied: 72 m/ha for pipes and pumping capacity of 200 1/min for deep-well stations. Also, the existing KIP standards were proposed for the detailed engineering design of facilities shown in Drawing 8 (water taps) and Drawing 9 (deep-well stations).

(b) MCKs (Public Washing/Bathing/Toilet Units)

(i) To rehabilitate/reactivate 3 existing MCKs now out of operation; to increase the attractiveness of MCKs by providing of 24 existing MCKs with low-cost roofing structures as well as with appropriate water supply from the public distribution network.

(ii)   To establish an additional 31 MCKs with a total of 248
         toilet units, to serve the population without any toilet
         system in their houses (See tabulation in Annex 22). The
         need for new MCKs, after rehabilitation and extension of
         existing ones, was calculated on the basis of 400 to 800
         persons per MCK (50 to 100 persons per toilet unit),
         depending on the area and population density of each
         particular kampung. In the calculation, the following
         factors were taken into account:

   (a)   The average number of persons currently using MCKs
         in operation is 53 families, i.e., 370 persons
         (See Annex 26).

   (b)   With an increase in living standards more families
         will install toilets in their houses.

   (c)   People without toilets of their own are spread
         within the kampungs and their readiness to go to
         an MCK (according to the survey results) is
         generally limited to a distance of 50 to 100 m.

   (d)   There is limited availability of land within highly
         populated kampungs.

In order to have sufficient flexibility for location of MCKs, with
regard to the actual needs and land availability, three types of MCK were
considered: with 4, 8, or 12 toilet units, and the existing KIP standards were
proposed for the detailed engineering design (See Drawings 10 and 11).

   (c) Leaching Pits

      (i)   To implement in two kampungs, one in Setiabudi the other
              in Tebet, a pilot project consisting of 40 leaching pits
              of new design (larger capacity), to provide a basis for
              new regulations under consideration concerning toilet
              systems in houses as well as for the construction of the
              remaining 3,000 leaching pits in the area to be built
              at the cost of the householders themselves (See Annex
              23). The typical design of leaching pits now built
              by KIP shown in Drawing 13, is judged to be of insufficient
              capacity.

      (ii)   To provide technical and financial incentives to families
              to build leaching pits including technical assistance,
              easy credit, or subsidies. It was suggested that the
              extent of subsidies and other financial incentives should
              be determined, in each particular case, in consultation
              with a committee consisting of representatives of local
              authorities and the population.

(d) <u>Additional Excreta Management Measures</u>

In order to achieve the objective of total excreta management, **it was further** recommended that an ordinance needed to suit the new situation **should be** based on the following principles:

(i) All buildings close enough to sewers must be forced to connect and to pay a connection fee. For houses of poor people, the government could permit payment by installments over a long period.

(ii) All houses in kampungs not in (i) above must be required to have adequate leaching pits and to keep them pumped out as frequently as necessary. The Government is expected to provide standards for acceptable pit design and requirements on desludging, and to help poor families with payment where the required frequency of pumping is over a predetermined limit. For construction of pits two approaches could be considered. The Government builds the facility and collects the money, or the families are required to build the facility with easy Government credit.

(iii) All persons not having their own toilets, nor access to the toilets of others, should **be encouraged** to use MCKs to be furnished by the Government.

(e) <u>Desludging</u>

(i) To provide additional desludging equipment: 5 vacuum trucks of 2 m^3, 1 vacuum truck of 6 m^3 and 2 motorized vacuum trailers of 0.5 m^3 capacity and a width of 1 m, able to enter narrow footpaths and, thus, to serve houses in zones inaccessible for ordinary vacuum trucks. The sludge collected by vacuum trailers could be then received by a vacuum truck on the closest road. The City Cleansing Department has already provided 6 units of such small desludging trailers (weight 300 kg empty, 800 kg loaded). They have been manufactured locally on the basis of a Japanese design and modified for kampung conditions. Their practical use was, however, hampered by certain technical problems (manual moving to sites of their effective use, operation, coordination with vacuum trucks for sludge transfer, etc.) requiring additional conceptual modifications. Besides, some financial and operational aspects (desludging fee and system of its collections) were still, at the time of the study, under discussion with local authorities who are supposed to manage and operate this equipment under a long-term contract with the City Cleansing Department.

(ii)   To set up a desludging service exclusively for poor kampungs, in order to avoid omitting of poor people by desludging crews who give priority to richer people, expecting higher extra-income.

(iii)   To establish a transfer and thickening station in the area with the following two main purposes:

    (a)   Reducing the volume of sludge by discharging supernatant water into the proposed sewer system.

    (b)   Transfering the remaining thickened sludge to tanker trucks for more economic transport to the existing sludge treatment plant located 20 km away from the area.

As shown in Drawing 14, the sludge transfer and thickening station basically consists of two open sedimentation tanks. With alternating operation, one tank will always be ready for receiving sludge collected from two vacuum trucks. The sludge will be allowed to settle during the night. The following morning, the supernatant, assumed to be at least 50 percent of the volume, will be released by gravity to the nearby sewer. The thickened sludge or underflow will be pumped to a tanker truck for transport to sludge disposal sites. After removal of all sludge, the tank will be rinsed clean and will be ready for receiving sludge on the morning of the third day.

The station is dimensioned to serve the project area with alternating operation of the two tanks. This will provide ample flexibility for operation and will not require any night shift work. However, with an adjusted timetable for operating (removing the thickened sludge in the early morning hours so that both tanks are empty at about 9 am), the capacity of the transfer and thickening station could be doubled. It could, due to its central location, easily serve also nearby kecamatans (Menteng and Taman Abang), assuming an early morning shift for operators and tanker drivers is organized. In this way, the per capita cost of desludging would be reduced.

(f)   Surface Drains

(i)   To repair 176.8 km of damaged drains, comprising 110 km in kampungs and 66.8 km in non-kampung areas, and

(ii)   To establish 24.3 km of new drains, including 10 km in kampungs and 14.3 km in non-kampung areas, according to the survey results (See Annex 24). Typical profiles of drains to be constructed along the streets and footpaths, as recommended for detailed engineering design, are shown in Drawing 15 and 16, respectively.

A schematic drawing illustrating all the above improvements is shown in Drawing 17.

(g)   Operation, Maintenance, and Monitoring

This includes provision of a program for operation and maintenance and for monitoring of the above facilities, including the preparation of appropriate manuals and guidelines, so that the facilities will be accepted and used by the community to achieve their intended primary purpose of helping to improve environmental conditions in the poor areas.

The Jakarta Sewerage and Sanitation Project, while it has completed the sanitation survey described here, and has prepared preliminary criteria for the needed improvements, has not as yet actually constructed these improvements (construction is expected to start in 1983). Therefore guidelines for operation and maintenance and for continuing monitoring have not yet been prepared. These guidelines will be prepared as part of the new work program. At this time preliminary concepts on operation and maintenance and on monitoring are described in Annex 28, Section 7.

The importance of continuing monitoring, following completion of construction, can scarcely be overemphasized. Very few if any of the many urban slum improvement projects undertaken in developing countries over the past two decades have provided for such monitoring. As a result the facilities often tend to lose their value because they are not properly managed including insufficient attention to repairs and to administration and management. Through periodic monitoring it should be possible to correct this problem and thus greatly enhance the value of the investments in the facilities provided. Such a monitoring-cum-physical inspection program has been proposed for Surabaya (Reference 3).

It is planned to establish periodic monitoring for the Setiabudi - Tebet area following completion of the sanitation improvement project. For this purpose a "Manual of Guidelines for Periodic Monitoring of Community Sanitation" will be prepared within the implementation phase of the sanitation project.

5.2 COST ESTIMATES

(a) Capital Costs

The cost estimates were based on unit costs experienced in previous similar projects in the country updated to the present (piped distribution network, deep-well stations, public water taps, communal sanitation facilities, leaching pits, drains), on current local market prices (acquisition of land for communal sanitation facilities, deep-well stations, public water taps and sludge transfer and thickening station), or on international market prices (pipes, pumps and desludging equipment, including transfer and thickening station).

Because the proposals for sanitation improvements were based on the extrapolation of the sample survey's results, which of course implies possible needs for adjustments and more engineering work in the implementation phase the basic cost estimates (land excluded) were increased by 25% for physical contingencies and 15% for engineering. In addition, for the entire 3-year construction period considered, an average increase of cost of 20% has been considered for price contingencies (land included), based on an annual inflation rate of 10% expected in the country.

The total capital cost of the proposed improvements was estimated at 2,546.4 million rupiah (equivalent to US $ 3.92 million)[1]. A summary of cost estimates is presented in Annex 25 and its breakdown into local and foreign components in Annex 26.

(b) Operation and Maintenance

The operation and maintenance costs of the proposed new facilities were calculated partly on the basis of updated historical data for similar facilities in operation (public water taps, deep-well stations, communal sanitation facilities, drains), and partly on the basis of estimates for different cost components, such as labor cost, material, spare parts, energy, etc., according to the current local or international market situation (desludging services, including transfer and thickening station). For deep-well stations, the energy cost for water pumping represented over 85% of the total costs, but the operation and maintenance costs of other facilities consisted essentially of salaries for operators and caretakers (communal sanitation facilities - over 75%, desludging services - about 70%, drains about 90%). In some kampungs, however, where the cleaning and maintenance of communal sanitation facilities are carried out by the self-help of users, these costs are limited to desludging septic tanks, water, and electricity supply. It was therefore recommended to follow and extend the policy of self-help to all kampungs in order to reduce significantly the operation and maintenance costs of these facilities.

The operation and maintenance costs, estimated for the full operation of each facility, represent about 5% of capital costs (without land) for water supply, about 10% for communal sanitation facilities and desludging services, and 0.5% for surface drains. For individual leaching pits, no cost on the part of any government agency was associated with operation and maintenance. However, in some cases, for instance for poor families, particularly if the leaching pits are flooded, Government subsidies for repairing or desludging were recommended, adopting a similar approach to that suggested for construction.

The implementation of the proposed improvements was scheduled for the subsequent three-year period (See Annex 27). The time schedule was patterned on the KIP procurement system which has been used successfully for over 12 years in similar KIP projects, with financial assistance from the World Bank.

5.3 IMPLEMENTATION PROGRAM

The implementation program of the proposed sanitation project should include the following:

(a) Detailed engineering for all sanitation facilities to be constructed or rehabilitated to the extent this work has not been done in the feasibility study. If design standards for the facilities are already available and are also in practical use and accepted by the community, these should be applied since they simplify the work program including detailed engineering and operation and maintenance. If such standards are not yet available, technologies

---

1/ Exchange rate: US $ = 650 rupiah (May 1982).

mentioned in the World Bank Publication "Appropriate Technology for Water Supply and Sanitation, A Planning and Design Manual", Vol. 2 could be used, provided the community will accept them. In such cases, appropriate adjustments of standard designs to the location of particular facilities would be necessary.

(b) Setting up the implementation schedule and contract packages, preparing tender documents for bidding, evaluation of bids submitted, negotiations and awarding of contracts. The scheduling and packaging should be based on the following principles:

(i) All sanitation facilities in a particular area should be built, to the maximum possible extent, in the same period in order to avoid long-term distrubance of the population concerned.

(ii) Preparation of the construction packages for an area or group of areas, depending on the volume of work to be implemented.

(iii) Rehabilitation/repairing of existing facilities should be given priority.

(iv) Appropriate procurement procedures (international or local competitive bidding) should be followed, depending on the kind of supply and services as well as on financing sources (financing agencies) involved.

(c) Supervision during construction including: preparing detailed work schedules, checking the ordering and supply of materials and equipment, supervising the execution and progress of construction work and related tests, reviewing invoices of contractors and suppliers, commissioning and handing over of the completed works.

(d) A detailed review of the existing system of operation, maintenance, and monitoring for each particular sanitation component and proposals for their improvement, to ensure that the rehabilitated/reactivated and newly established sanitation facilities will operate appropriately. In this connection , theoretical and practical on-the-job training of technical and administrative staff engaged for the above activities should be carried out.

(e) Preparation and execution of an educational campaign for the population in the area with the aim of obtaining their support and acceptance for the new sanitation facilities.

The above work can be done with or without assistance of consultants, depending on the capacity, experience, and ability of the government agencies involved.

For the implementation of the Jakarta sanitation project, the assistance of competent engineering consultants was recommended. For illustration, Terms of Reference for the needed consulting services, which cover practically all aspects of the implementation program, are presented in Annex 28. Examination of the full text of this document is the best way to demonstrate the multiplicity and complexity of the tasks to be accomplished in the implementation phase. It demonstrates the necessity for an efficient permanent coordination of all activities during implementation as well as in the post-implementation period of the project. Consequently, it can serve as a basis for the setting up of implementation programs for similar projects.

CHAPTER 6

PLANNING SURVEY

The Jakarta experience as described above shows that four main facotrs may influence the planning of the survey, as follows:

(a) Extent of the area and size of the population to be covered by the study.

(b) Diversity of settlements/households to be surveyed.

(c) Availability and ability of surveyors.

(d) Budget available for the survey.

The greater the population and the larger the area to be covered, the more households have to be interviewed in order to obtain the expected reliability and representativeness of data collected. For the same reason, more sample survey zones are required, if the settlement and family patterns are very different. If the surveyors available are of limited capability, more time will be necessary for carrying out the survey than for more qualified surveyors. In the case of budget limitations, a compromise must be found in order to achieve an acceptable degree of reliability of the data base to permit the planning of appropriate sanitation improvements and priorities.

The number of people/households within the project area is a basic criterion for identifying the magnitude of the survey, not only for house sanitation facilities, but also for public facilities (water taps, communal sanitation facilities, desludging services, solid wastes disposal , surface drains), since these are also related directly or indirectly to the size of population. Thus, all relevant parameters for planning the survey, such as survey extent, staffing, scheduling, logistical support and costs, will refer to this criterion. To facilitate the planning of particular surveys, the calculation of these parameters is illustrated in Annex 29 and 30. More details are given in the following discussion.

6.1 SURVEY EXTENT

Generally, it is desirable to carry out as large a sample survey as possible. The larger the data basis is the higher its reliability for planning purposes. On the other hand, manpower and financial constraints usually place limits on this aim. In each particular case, therefore, the survey extent (proportion of the population to be surveyed) will finally be determined at a point between the "desirable" and the "possible". Neverthe-less, on the basis of the Jakarta experience, the following paramaters are recommended as guidelines for defining the extent of the sample survey in relation to the total population of the area (See Annex 29):

Total Population within the Area	Population to be Surveyed	Percent of the Total Population
25,000	2,500	10.0
50,000	4,500	9.0
100,000	8,000	8.0
200,000	12,000	6.0
300,000	15,000	5.0
400,000	18,000	4.5
500,000	21,000	4.2
600,000	24,000	4.0
700,000	26,000	3.7
800,000	28,000	3.5
900,000	29,000	3.2
1,000,000	30,000	3.0

It is assumed that the diversity of settlements/households does not increase proportionately with the total size of population and that, consequently, various categories of sample survey zones could be simply extrapolated to a higher number of similar zones, i.e., to a larger total population. On the contrary, the smaller the area, very probably, the higher should be the proportion of the population to be surveyed, in order to cover all categories of population involved.

In the Jakarta project, which covered an area with a population of about 500,000, some 12,235 people were surveyed (1,826 households with an average of 6.7 persons per house), i.e., 2.5% of the total area population, during one month (20 fieldwork days) only. Had the survey lasted two months, the recommended ratio of 21,000 people could then have been achieved easily. But, taking into account the relatively high degree of extrapolation possibilities, it is believed that even in these conditions, the data collected provided an adequate basis for quantifying the sanitation needs and identifying the improvements and priorities.

6.2 STAFFING

Guidelines for estimating staffing requirements for planning a monitoring survey are discussed below.

6.2.1 Professional Personnel

(a) Study team leader (project manager):

To be in charge of preparation, detailed planning, organization and management of the survey as well as of field data analysis and report preparation. His main duties would be:

(i) To collect all relevant data and documentation related to the area to be surveyed (See Section 2.1).

(ii) To divide the project area into different sub-areas, to classify them according to prevailing housing patterns (family income groups), population density, flood risks and other criteria; thereafter, to select and delineate the sample survey zones (Section 2.2).

(iii) To prepare and test the survey questionnaires for different sanitation components, in accordance with the survey objectives and particular local circumstances (Section 2.3).

(iv) To set up the survey team and organize and manage its training (Section 2.4).

(v) To set up the survey organization chart with a clear delineation of the responsibilities and duties of the different categories of personnel involved in the execution of the survey, as well as a general plan for carrying out the survey (Section 3.1).

(vi) To prepare daily detailed plans for the fieldwork and to provide all appropriate documents, working materials and equipment, as well as suitable rooms for meetings and office work and transportation logistics (Section 3.1).

(vii) To cooperate closely with local authorities during the survey, in order to obtain their support for the field-work as well as for the proposed improvement measures (Section 3.3).

(viii) To arrange for processing of survey data (Section 4.1).

(ix) To evaluate the survey's results and to propose appropriate improvements, as well as the related implementation program (Section 4.2 to 5.3).

The study team leader should be a graduate in socio-economics or in sanitary engineering. Because of his many complex and varied duties, involving large numbers of personal contacts and sensitive negotiations, priority must be given to a specialist with considerable organizational and managerial capacities and dedication. Because his managerial work requires many contacts with central and local authorities, and with the various committee involved in the survey, in addition to his daily intensive work with the surveyors, the team leader must have great abilities in public and human relations. While for particular aspects of the project, he could easily be advised/supported by appropriate specialists, e.g. by sanitary engineers for technical and technological aspects and by public health specialists for health problems, he must himself be responsible for management, coordination, and public relations.

(b)   Sanitary engineer:

The main duty of the sanitary engineer is to assist the study team leader (project manager) in the technological aspects of the project, especially in the following fields of activities:

(i)   Collection and interpretation of geological, topo-graphical, and hydrogeological maps and aerial photo-graphs covering the project area.

(ii)   Collection and evaluation of plans and standard designs of existing and proposed sanitation facilities and equipment, such as piped water supply systems (including wells and other water sources, mains and distribution network, and water taps), household and public sanitation facilities, solid wastes disposal facilities, desludging systems and equipment, etc.

(iii)   Classification of sub-areas and selection of sample survey zones.

(iv)   Determination of questions and possible answers in the phase of questionnaire preparation related to particular technical parameters of facilities to be surveyed.

(v)   Training of surveyors.

(vi)   Evaluation of survey data.

(vii)   Idendification and quantification of needed improvements, including acceptable technologies and technical designs of proposed facilities as well as improvements of opera-tion and maintenance systems.

It is understood that both the study team leader and his assistant, the sanitary engineer, should be engaged full-time for the entire period of the project. If the population of the project area exceeds 500,000, it is recommended that an additional sanitary engineer be engaged to help in identi-fying and quantifying improvement needs as well as in detailed design of proposed facilities.

(c)   Surveyors:

As already mentioned (Section 2.4), the selection of surveyors is an important issue because the success of the work depends, to a large extent, on their personal qualities, professional motivation, and knowledge of the living conditions of the people to be surveyed. University students or practic-ing graduates of sanitary engineering appear to be those with the most suitable background for surveyors. As explained in Section 3.2, they should work in the field in 2-man working groups.

The total number of working groups depends on the population to be surveyed based on a ratio of about 10 households to be interviewed (i.e. about 60 persons to be interviewed) per group per day as was the experience in Jakarta (see Section 3.2). Estimates of the required number of working groups with related field-working days are presented in Annex 30. As can be seen, about 16.7 field-working days per 2-man group could be calculated for each 1,000 persons to be surveyed.

With fewer surveyors more field days are required and vice-versa. The Jakarta experience, however, shows that both these variables have limits. It is not easy to find a high number of qualified surveyors and, even if available, the management and coordination of their fieldwork might be very complicated, due to the reduced mobility of the team within the city, particularly in overcrowded slum urban areas. On the other hand, the efficiency of the surveyor's work may go down, if he must work over a very prolonged period. This is particularly true if no special incentives are granted to them. According to the Jakarta experience, it is recommended not to exceed the number of 30 surveyors (15 two-man groups) and not to exceed a 2-month period for the effective field-work.

The tabulation in Annex 30 is designed to help determine the appropriate number of surveyors and appropriate survey period. In addition to actual field-working days, from 2 to 3 days have to be added for training of surveyors. For a team of 6 surveyors or less, the training can be reduced to 2 days.

(d) Coordinator:

A coordinator is not needed when only a few surveyors are used; their work in the field can easily be coordinated by one of them, preferably the most experienced surveyor. However, if the team consists of more than 10 surveyors (5 two-man groups), a coordinator will be needed, preferably a senior sanitary engineer. If the team will consist of more than 20 surveyors (10 two-man groups) two coordinators would be required. The coordinator should be responsible for the following:

(i) To distribute the questionnaires with appropriate documents (maps, sketches) and equipment received from the study team leader (project manager), to each two-man group before departure to the survey area.

(ii) To supervise the work of surveyors and to help them when problems arise and advice is needed, e.g., coordination of team transportation within the area, distribution of tasks in the case of unexpected events (illness of surveyor, road accidents, etc.); to collect the filled-in questionnaires and to give them to the study team leader for subsequent data processing.

The role of the coordinator is particularly useful, even indispensable, if an expatriate specialist is appointed for managing the study, who does not speak the local language. In such a case, the coordinator functions also as a translator and intermediate between the surveyors and expatriate staff.

6.2.2 Supporting Staff

(a) Draftsman

Because a lot of drawings are needed for the survey as well as for the final report, a full-time draftman is required during the entire. study period. If the study area will cover more than 500,000 people, two draftsmen will probably be needed.

(b) Secretary

A full-time secretary is required for the entire period of the study, no matter the size of the survey. For writing the final report of the project, however, it may be necessary to engage an additional part-time secretary to manage the typing load.

6.2.3 Summarized Estimates of Staffing Needs

The detailed estimates of staffing requirements as related to survey size are shown in Annex 29. A summary presentation is given in Table 1.

Table 1: Staffing Requirements for Monitoring Survey

Population to be surveyed	Managerial Staff		Surveying Staff		Supporting Staff		Total	
	man	man/day	man	man/day	man	man/day	man	man/day
2,500	2	126	4	92	2	126	8	334
4,500	2	126	8	176	2	126	12	428
8,000	2	126	15	330	2	126	19	582
12,000	2	146	15	480	2	146	19	772
15,000	2	168	15	585	2	168	19	921
18,000	2	168	17	680	2	168	21	1,016
21,000	2	188	19	798	2	188	23	1,174
24,000	3	209	21	903	3	209	27	1,321
26,000	3	235	24	1,032	3	235	30	1,502
28,000	3	235	26	1,092	3	235	32	1,562
29,000	3	235	26	1,118	3	235	32	1,588
30,000	3	235	26	1,148	3	235	32	1,618

6.3 SCHEDULING

The scheduling of the monitoring survey, including preparatory work, collection and evaluation of data, and writing the report, depends primarily on the time required for fieldwork. The tabulation of the field-working days,

presented in Annex 30, indicates possible periods for this work as related
to the number of surveyors considered. The total time of the fieldwork, however, as mentioned above, should not exceed two months, to avoid a decrease in
efficiency.

The preparatory work, including planning of the study, collection
of background documents and information related to the study area, selection of
sample survey zones, preparation and testing of questionnaires, and setting
up and training of survey team, requires 1.0 to 1.5 months, depending on the
size of the area to be surveyed. About the same period of time is needed for
evaluation of survey results, identification of improvement measures, and
setting up an implementation program, assuming the processing of survey data
and their extrapolation to the entire project area is done step by step during
the survey, so that all the computed data will be available shortly after the
last sample survey is finished. This will permit starting the preliminary
design of sanitation components during the survey, as soon as the first survey
findings are available and evaluated.

On the basis of the above assumptions, estimates were made for the
scheduling for various sizes of surveys. The results are presented in Annex 29.
As shown, the time schedules for the entire study vary from 3 to 5 months, with
the time required for fieldwork from about 1.0 to 2.0 months. Generally, the
overall sanitation study may be depicted as follows:

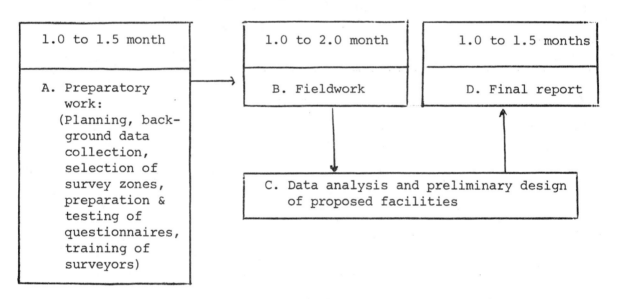

However, if data and documents required for the survey are not
available (or partly only) and consequently more work will be necessary in the
preparatory phase (see Section 2.1), the time schedule has to be adequately
extended and the survey budget appropriately adapted.

Annex 8 explains the scheduling used for the Jakarta survey, which
illustrates the concepts noted above. As can be seen, the time for preparatory
work was limited to an one-month period, with a one-month period for the actual
field work. This was possible due to the availability of all data and documents
required for preparing the survey (see Section 2.1) as well as to a good
cooperation with local authorities and people concerned during the whole
period of the survey (see Section 3.3). Though normally about 1.5 months for
preparatory work and about 2 months for field investigations would be required

for such a project, the above favorable conditions permitted taking advantage
of the computer - supported approach. Thus in the short time a reliable data
base for planning purposes was established (see Section 1.2).

6.4 EQUIPMENT AND SUPPLIES

As mentioned above (Section 3.1), in addition to an appropriate
quantity of questionnaires with maps and sketches, the following equipment
and logistical support is required for the survey:

(a) Adequate number of vehicles with drivers:

(i) One car for study team leader.

(ii) One car for each sanitary engineer.

(iii) One mini-bus per 10 persons for survey team.

Because the study duration is quite limited, it is preferable to
rent these vehicles for the anticipated period, with an appropriate provision
for fuel, services and for insurance if these expenses are not included in
the rent contract.

(b) Fieldwork equipment:

(i) Water pressure gauges (one for each group) for measuring
the pressure of water in water taps.

(ii) Acoustic or light level gauges (one for 4 groups) for
measuring the water level in wells; this equipment
could be kept by the coordinator and given to the surveyors always as needed.

(iii) Metallic tapes (one for each surveyor) for measuring
surface drains, wells, and other purposes.

(iv) Auxiliary field materials, such as desk pads for questionnaires, maps and sketches, pens/pencils and erasers
(one set for each surveyor). If the survey is carried
out in the rainy season, one umbrella, at least, for
each group should also be provided.

(c) One meeting room for daily meetings with the survey team,
as well as one office room with appropriate secretary and drafting services
for management of the survey. Because the sanitation aspects of the city are
usually the responsibility of the Municipality, these facilities and services
will be provided usually by the Municipality. If not, appropriate provision
should be budgeted for these purposes.

6.5 COSTS

The budget for the monitoring survey should comprise the following
components:

(a) Salaries, living allowances, international travel and related costs for expatriate staff (if required) according to the engagement contract.

(b) Salaries, including all social and other charges, for the local staff, according to the local regulations.

(c) Local transportation costs including rents for vehicles with drivers, as well as fuel, insurance, repairing and maintenance services, to the extent these are not included in the contract with the charter.

(d) Cost of equipment and supplies, to the extent these are not available free of charge (e.g., water pressure and water level gauges, office equipment and furniture, etc.) by the Municipality or other agencies.

(e) Special allowances/incentives for surveyors. These are particularly recommended if the field investigations would last more than two weeks.

(f) Cost of data processing.

(g) Miscellaneous (reproduction of questionnaires, maps, sketches and other documents, printing and binding of report, communications, etc.).

It is understood that the cost estimates will vary from one country to another, according to the actual local cost levels for different items. For illustration only, the total cost related to the Jakarta sanitation survey, including the assistance of expatriate staff (full-time project manager and part-time sanitary engineer) was about US $ 100,000, i.e., 2.5% of the total investment costs of the proposed sanitation improvements, or $ 0.20 per capita in relation to the total population of the project area.

CHAPTER 7

SUMMARY AND CONCLUSIONS

(a)   The Jakarta sewerage and sanitation project, now being implemented by the Government of Indonesia with assistance from the World Bank, is a pioneering one in that it has the objective of achieving comprehensive environmental clean-up in a pilot or demonstration area of Jakarta (the Setiabudi-Lebet kecamatans) with a population of about 500,000. In addition to provision of Jakarta's first program of sewerage, the project includes construction of a variety of sanitation improvement measures, mostly in the poor people or kampung zones, including (i) public water taps, (ii) surface drainage improvements, (iii) improvements in the system of individual leaching pits used by most poor people homes for disposal of excreta, including improvements in pit desludging services, (iv) provision of public washing/bath-ing/toilet units, and (v) through a complementary project, improvements in solid waste management.

(b)   A major problem in planning the program of sanitation improvements was lack of information on the current status of sanitation facilities in the project area, namely information on the existing facilities and their adequacy, and on the gaps to be filled in order to achieve a minimum desired level of environmental cleanliness throughout the project area. While a variety of such facilities had been installed in the project area over the past 15 years, mainly through the Jakarta Kampung Improvement Program, factual information was lacking on the extent of their adequacy/inadequacy due to the lack of suitable monitoring of the use, condition, and impacts of the facilities. It was recognized, while several socio-economic surveys had been made, none of these produced the "hard data" needed to permit an engineer to design a specific program of improvement measures with the assurance that the improvements would indeed fill the gaps if operated and managed properly, and with the assurance that the facilities, once built, would be properly operated and managed.

(c)   The need for this type of engineering monitoring surveys for quantify-ing sanitation gaps at Jakarta was first noted in 1976, as part of the UNDP/WHO report on a comprehensive plan of sewerage and sanitation for metropolitan Jakarta. The ongoing Jakarta Sewerage and Sanitation Project (JSSP) represents the first step in implementing the concept of sewerage-cum-sanitation proposed in the master plan.

(d)   Making use of the WHO/UNDP sanitation studies and several subsequent World Bank appraisals of sanitation needs at Jakarta, a unique sanitation survey methodology combining monitoring of sanitation facilities and socio/cultural/economic factors affecting their use was designed and applied to the Setiabudi-Tebet project area. The present report:

(i)   Describes how the survey was done at Jakarta, on a step-by-step basis, including preliminary assessment of scope of work involved, design of questionnaires to be used by field team members to obtain relevant data directly usable for computerized analysis, selection of appropriate sample areas, training of the field team members, planning of the field work, execution of the field work, and collation and analysis of the data for the purpose of delineating and quantifying the needs for sanitation measures, all so presented to indicate clearly the concept and rationale for the overall survey plan and for each step, with the actual data from the Setiabudi-Tebet survey used to illustrate the overall and step-by-step processes in planning and conducting the survey. (Chapter 2,3 and 4)

(ii) Describes how the results of the survey were utilized for preliminary design of the needed improvements. (Chapter 5)

(iii) Based on the Setiabudi Tebet experience, presents guidelines to assist government officials in planning similar monitoring surveys in other urban areas in developing countries, including criteria on needs for personnel and for equipment and supplies together with time and budget requirements. (Chapter 6).

(e)   Because of increasing recognition in developing countries of the critical need for sanitation improvement programs in urban poor people areas and because planning and implementation of such programs is hardly possible without first clearly delineating the needed facilities, including preliminary design criteria, so that a specific minimum cost package of improvements measures can be planned with assurance that the gaps will be filled, it is anticipated that the methodology on sanitation monitoring surveys used at Jakarta could find wide application in the developing countries.

(f)   An additional important need, not yet believed to be implemented anywhere, is for a continuing minimum-cost program of continuing monitoring of community sanitation facilities, to check on needs for repairs on whether the facilities are properly administered, maintained, and managed, and on their socio-economic acceptance including impact in increasing public desires for improved sanitation and willingness to pay for it. This is simple work but simply isn't done, and it is critically essential for getting a meaningful return on the investment in the facilities. It would produce the hard data not only needed for effective repairs but equally needed for proving the value of the facilities and for improving fundamental planning and design concepts. It is assumed that such periodic monitoring will be implemented as part of the Jakarta Sewerage and Sanitation Project. Annex 28 contains brief terms of reference for a monitoring survey and Annexes 29 and 30 present staff and field-working days estimates for such a survey.

ANNEXES

ANNEX 1

JAKARTA SEWERAGE AND SANITATION PROJECT (JSSP)
SANITATION SURVEY

QUESTIONNAIRE A   - HOUSE SANITATION FACILITIES

A) INTRODUCTION

B) QUESTIONNAIRE

C) EXPLANATORY NOTES

D) FILLED-IN QUESTIONNAIRE IN LOCAL LANGUAGE
    (FIRST 3 PAGES)

INTRODUCTION

1.	The attached questionnaire, used in the Jakarta sanitation survey serves as an example of how to build up such a document (formulating of questions and possible coded answers) in order to facilitate the collection and processing of data on those sanitation components/ parameters which are relevant for the planning of appropriate sanitation improvements and priorities. As can be seen, the selected parameters include not only the physical house sanitation facilities (water supply, toilet systems, leaching pits, septic tanks, etc.), but also - if such facilities do not exist on the plot - the acceptance and use of public ones, as well as the readiness and willingness of the people concerned to participate in their construction, operation and maintenance. In addition, they also include housing and family patterns, some aspects of people's hygienic, institutional, managerial and other aspects related to the project.

It is understood that the quite comprehensive questionnaire was required, since the sanitation survey was focused, among other things, on the facilities built in the last 12 years within the Kampung Improvement Program (KIP), particularly on their physical condition, operation and maintenance, acceptance and effective use by the intended beneficiaries. So, the sanitation survey was practically combined, to a large extent, with monitoring in order to provide through a critical analysis of existing KIP facilities, an appropriate data base for new design of these facilities and adequate improvements of their operation and maintenance systems. In this connection, for example, the desludging aspects were dealt with separately for septic tanks and leaching pits, since particular attention had to be paid to leaching pits in poor highly populated areas. Further, important data and information were needed for a proposed educational campaign to be organized (before implementation of the project) for the population within the area, with the aim of obtaining their support for and acceptance of the new sanitation facilities considered, as well as to inform them about respect for hygienic principles in using water, toilets and other sanitation services.

It is understood that the sanitation parameters could vary from country to country and from city to city, and therefore, the shaping of the questionnaire will vary accordingly, depending on the project being considered. For some projects, probably only the most relevant data (as mentioned in Section 4.3) will be necessary; for other projects maybe, even more parameters (e.g. health status of the population, communal waste disposals, etc.) will be required. Therefore, at the beginning of each study, all relevant sanitation parameters to be surveyed must first be selected according to the assigned objectives of the project, and thereafter, an appropriate hierarchy of questions must be established.

In any case, an attempt should be made to identify key questions that are essential for the survey (identification of main sanitation problems and improvement needs, information about acceptance of considered new technologies or improvements of existing ones by the community), in order that the disturbance to the people interviewed is reduced, but the survey target achieved.

2.   The persons to be interviewed are preferably the family head and his wife; in the case of their absence, of course, any other adult family member could be asked for information, as far as he is in the position to give it. If not, an arrangement could be made for a later visit or to interview another family within the sample survey zone.

3.   It is a matter of course that a large amount of data related to house sanitation problems has to be collected from appropriate authorities, such as water rates and house connection fees for piped water supply, desludging fees, systems and methods, soil permeability within the project area, etc. On the other hand, much of the data and information obtained from households is very useful, even essential for planning, the operation and maintenance of public sanitation facilities (public water taps, communal toilets, etc.), for example: the reasons why these facilities are not used by the people concerned, what are the preconditions for their acceptance by intended benefi- ciaries, for participating in their construction, operation and main- tenance, etc. In other words, the household interviews have to be considered as one of the important information sources concerning house sanitation aspects, but at the same time as an integrated part of data collection related to the overall sanitation environmental problems within a project area.

JSSP - SANITATION SURVEY

A - House Sanitation Facilities

Surveyor(s): .... Date ....

Kecamatan: .... Kelurahan ....

Kampung: .... Street .... House No. ......

Owner: ....

Person(s) interviewed: ....

House/Plot - Sketch   (Scale 1 : ....)

R = Room	HP = Hand pump
K = Kitchen	DW = Dug well
B = Bathing	SP = Standpipe on plot
W = Washing	◉ = Water tap (if piped supply)
T = Toilet	WT = Water tank
LP = Leaching pit	D = Drains, ditches
ST = Septic tank	SW = Solid wastes

Ref. No.: .../...

1. <u>ZONE</u> (K = Kampung, NK = Non-Kampung)

Card - 1

1 Zone
| 1 | | |

K	-	1	Kampung	1	Duku Setiabudi	1
		2		5	Karet Belakang II	2
		3		6	Karet Pedurenan	3
		4		7	Kuningan I	4
		5		9	Kuningan II	5
		6		10	Kuningan III	6
		7		11	Karet Sawah	7
		8		13	Kawi Gembira	8
		9		14	Menteng Wadas I	9
		10		16	Menteng Atas	10
		11		17	Menteng Rw. Panjang	11
		12		18	Kebon Obat	12
		13		21S	Warung Pedok	13
		14		22	Warung Pedok II	14
		15		24W	Manggarai Barat/Timur	15
		16		25	Bali Matraman	16
		17		27	Bukit Duri Selatan	17
		18		28E	Melayu Kecil/Bukit Duri	18
		19		32	Tebet Timur	19
		20		36	Kebon Baru	20

NK	-	1	Non-Kampung Area	9-A	21
		2		12-A	22
		3		22-B	23
		4		25-A	24
		5		38-B	25

<u>HOUSING & FAMILY</u>

2. <u>Access Way</u>

4
| |

Vehicular road	0
Paved path	1
Unpaved path	2

If path, <u>distance of house to vehicular road</u>:

5
| |

< 25 m	0
25 - 50 m	1
51 - 75 m	2
76 - 100 m	3
> 100 m	4

3. <u>Housing Pattern</u>

6
| |

Permanent (solid structure)	0
Semipermanent (solid foundation, temporary superstructure)	1
Temporary (bamboo-wood structure)	2
Transient (temporary structure, small size)	3

4. Number of <u>rooms</u> in the house

7
| | |

5 Number of <u>separate families</u> in the house

9
| |

6. The total number of <u>occupants</u> in the house

7. Number of <u>adults</u> (persons aged above 15)

8. <u>Children</u> aged under 5

        aged 5 - 15

9. Main <u>occupation</u> of the family head:

    - Trade   0
    - Handcrafts (.............................)   1
    - Farming, fishing   2
    - Worker   3
    - Administration, teaching   4
    - Military service, police   5
    - Other (.................................)   6
    - Unemployed   7

10. Does <u>the owner</u> live in the house?   No   0
                                                Yes   1

11. If rooms are rented, what is <u>the charge</u>
    <u>per room</u> per month?

              < Rp 10,000   0
    10,001 - Rp 20,000   1
    20,001 - Rp 40,000   2
    40,001 - Rp 60,000   3
              > Rp 60,000   4

12. Estimated <u>family income</u> per month

              < Rp 30,000   0
    30,001 - Rp 55,000   1
    55,001 - Rp 120,000   2
    120,001 - Rp 200,000   3
              > Rp 200,000   4

    <u>WATER SUPPLY</u>

13. How is <u>water obtained</u>?

    a) <u>for drinking/cooking</u>

        Metered home connection   0
        Single standpipe on plot   1
        Water taps/cistern, public (KIP)   2
                            shared   3
        Deep well (>8 m), public (KIP)   4
                          shared   5
                          private   6
        Hand pump, shallow well (<8 m), public   7
                                      private   8
                                      shared   9
        Dug well with bucket scoop, private   10
                                    shared   11
        Vendors   12
        Other sources (......................)   13

b) <u>for hygienic and other purposes</u>

☐

As above (0 - 6)	0
Hand pump, shallow well (<8 m), private	1
shared	2
Dug well with bucket scoop, private	3
shared	4
River	5
Otherwise (.........................)	6

14. If taken from shared/tap/cistern/deep well

☐

a) What is the <u>distance to the house</u>?

< 25 m	0
25 - 50 m	1
51 - 100 m	2
101 - 200 m	3
> 200 m	4

b) How many people share the shared tap? ....

15. If <u>piped supply</u>

☐

a) What is <u>the pressure</u>?

No water (Why? ....)	0
< 0.5 bar	1
0.5 - 1.0 bar	2
1.1 - 2.0 bar	3
2.1 - 4.0 bar	4
> 4.0 bar	5

b) What is the <u>frequency of low pressure</u> (0.5 bar)?

☐

sometimes (............... times per day)	0
often (............... times per day)	1

c) If no water, does the family use a <u>small cistern</u> or <u>jug</u> to store water during the times when good pressure is available?

☐

No	0
Yes - small cistern (capacity ... 1)	1
- jug (capacity .... 1)	2

16. <u>If water used from dug well:</u>

a) What is the actual water level in the well, measured from well's upper edge (.... m) and less the height of the edge from ground level (...... cm)

19

< 1 m	0
1 - 2 m	1
2 - 3 m	2
3 - 4 m	3
4 - 5 m	4
5 - 6 m	5
6 - 7 m	6
7 - 8 m	7
8 - 9 m	8
9 - 10 m	9
10 - 15 m	10
> 15 m	11

b) <u>Age</u> of well:

31

< 2 years	0
2 - 5 years	1
6 - 10 years	2
11 - 15 years	3
16 - 20 years	4
> 20 years	5

c) <u>Cost</u>: Rp ....

17. <u>If dug well</u>, what is the

32

<u>diameter</u>:

< 1 m	0
1 - 1.5 m	1
1.5 - 2 m	2
> 2 m	3

33

<u>casing</u>:

concrete, entirely	0
partly (...... m)	1
masonry, entirely	2
partly (...... m)	3
no casing	4

34

<u>depth</u>:

< 3 m	0
3 - 4 m	1
4 - 5 m	2
5 - 6 m	3
6 - 7 m	4
7 - 8 m	5
8 - 9 m	6
9 - 10 m	7
10 - 15 m	8
> 15 m	9

18. How far is the <u>shallow or dug well</u> from the <u>nearest leaching pit</u>?

35

< 5 m	0
5 - 7 m	1
8 - 10 m	2
> 10 m	3

19. How much water is <u>used</u> (rough estimates in
    1/day/family). Total:

    Thereof for:

    - drinking and cooking

    - personal washing, bathing

    - using in toilet (flush, cleaning and hand washing)

    - washing clothes and other purposes

20. If water from shallow well is used, is it
    <u>boiled</u>

    a) before <u>drinking</u>?

    | No | 0 |
    | Yes | 1 |

    b) before <u>brushing teeth</u>

    | No | 0 |
    | Yes | 1 |

21. How much is <u>paid for water</u> if supplied by
    vendors (in 18 1 = 1 tin)?

    | ❮ Rp 40/2 tins | 0 |
    | Rp 41 - Rp 50 | 1 |
    | Rp 51 - Rp 60 | 2 |
    | ❯ Rp 60 | 3 |

22. If not connected to piped water now, would
    the family be able and willing to connect
    and pay for it and if so, how much?

    a) <u>water rates:</u>

    | No | 0 |
    | Maximum: Rp 20/cu. m. | 1 |
    | Rp 25/cu. m. | 2 |
    | Rp 50/cu. m. | 3 |
    | Rp 75/cu. m. | 4 |
    | Rp 100/cu. m. and more | 5 |

    b) <u>for connection fee:</u>

    | No | 0 |
    | Maximum: Rp 50,000 | 1 |
    | Rp 75,000 | 2 |
    | Rp 100,000 | 3 |
    | Rp 125,000 | 4 |
    | Rp 150,000 and more | 5 |

TOILET

23. What <u>toilet system</u> is on the plot?   <u>57</u> ☐

   - None   0
   - Cistern flush WC   1
   - Pour flush squat plate   2
   - Ventilated latrine   3
   - Nonventilated latrine   4
   - Other (............................) 5

24. If no latrine,   <u>58</u> ☐

  a) <u>Where do the people defecate</u> themselves?

     - Neighbor's shared latrine   0
     - Public latrine (MCK)   1
     - Unoccupied land   2
     - River   3
     - Drains, ditches   4
     - Otherwise (.........................) 5

  b) <u>How far do they have to walk?</u>   <u>59</u> ☐

          < 25 m   0
     26 - 50 m   1
     51 - 100 m   2
    101 - 200 m   3
       > 200 m   4

25. <u>Why is MCK's latrine not used,</u> if no   <u>60</u> ☐
    latrine on the plot?

   - Too far from the house (.......... m) 0
   - Fee required is too high (Rp ....) 1
   - MCK's latrine not maintained appropriately 2
   - Lack of privacy   3
   - Other reasons (......................) 4

   If MCK is too far, how far are people   <u>61</u> ☐
   <u>willing to walk</u> to use it?

   Maximum: 25 m   0
             50 m   1
          100 m   2
          150 m   3
          200 m   4

   If fee is too high, how much are they
   willing and able <u>to pay</u> for use of
   toilet?

                                                       <u>62</u> ☐

   For <u>adult</u>: nothing   0
           maximum Rp 5   1
                 Rp 10   2
                 Rp 20 and more 3

for <u>children</u>: nothing   0
          maximum Rp 5   1
                  > Rp 5   2

26. <u>Is the liquid effluent from the toilets going</u>

- to sewer   0
- to septic tank with drain fields   1
- to septic tank with overflow to drains   2
- to leaching pit   3
- direct discharge to drains   4
- direct discharge to streams, rivers   5
- other (...........................)   6

27. What <u>material is used to clean the body</u>
after defecating (not for the hands, after
using latrine)?

- Water   0
- Toilet paper   1
- Other material (......................)   2

<u>SEPTIC TANK</u>

28. If toilet system is connected to <u>septic</u>
<u>tank</u>,

a) How long ago was it <u>installed</u>?

       < 2 years   0
    2 - 5 years   1
    6 - 10 years   2
  11 - 15 years   3
  16 - 20 years   4
     > 20 years   5

b) What was the <u>cost</u>?

        < Rp 100,000   0
100,001 - Rp 200,000   1
200,001 - Rp 400,000   2
400,001 - Rp 500,000   3
500,001 - Rp 600,000   4
        > Rp 600,000   5

c) <u>Desludging frequency:</u>

Never (not yet filled)   0
Once in every 6 months   1
             1 year   2
             1.5 years   3
             2 years   4
             3 years   5
             4 years   6
             5 years or more   7

d) <u>Why never desludged</u> even if filled?

- Lack of money   0
- Inaccessible to desludging truck   1
- Other reason (.....................)   2

e) <u>Who does the desludging</u>?

- DKI Cleansing Department   0
- Private company (.....................)   1
- Homeowner   2
- Other (.............................)   3

f) <u>Where is the sludge disposed of</u> if the house-owner is doing desludging?

- on open land   0
- into river   1
- to other place (...................)   2

g) What is the <u>desludging fee</u> effectively paid by the <u>family</u> (per cu. m.)?

      < Rp 1,000   0
Rp 1,001 - Rp 2,000   1
Rp 2,001 - Rp 3,000   2
      > Rp 3,000   3

LEACHING PIT

29. If <u>leaching pit</u> in use,

a) Where is the <u>location</u>?

- within the house   0
- outside the house   1

b) <u>The type</u>:

- unlined   0
- concrete/brick   1
- lined with bamboo mat   2
- other structure (.................)   3

c) <u>The size</u> (diameter):

     < 0.8 m   0
0.8 - 1.0 m   1
1.1 - 1.3 m   2
1.4 - 1.6 m   3
     > 1.6 m   4

d) <u>The depth:</u>   76 □

         < 3 m   0
   3 - 6 m   1
   7 - 10 m   2
     > 10 m   3

e) <u>The age:</u>   77 □

      < 2 years   0
   3 - 5 years   1
   6 - 10 years   2
  11 - 15 years   3
     > 15 years   4

f) <u>The construction cost</u> (rough estimate):   78 □

         < Rp 30,000   0
  Rp 30,001 - Rp 40,000   1
  Rp 40,001 - Rp 50,000   2
  Rp 50,001 - Rp 60,000   3
        > Rp 60,000   4

g) <u>Who has built it?</u>   79 □

  - Local craftsman   0
  - Self-made   1
  - Under KIP   2
  - Other (...........................) 3

CARD 2   1 Zone
              2 ☐☐

               4 ☐☐

h) <u>What is the ground water level?</u>

        < 1 m   0
   1 - 2 m   1
   2 - 3 m   2
   3 - 4 m   3
   4 - 5 m   4
   5 - 6 m   5
   6 - 7 m   6
   7 - 8 m   7
   8 - 9 m   8
   9 - 10 m   9
  10 - 15 m   10
     > 15 m   11

30. <u>How fast does the pit fill up?</u>   6 □

        < 6 months   0
  6 months - 12 months   1
  1 year - 2 years   2
  2 years - 5 years   3
     > 5 years   4

31. <u>Does the level in pit rise during rains?</u>

7

- Does not rise   0
- Rises to just below the slab   1
- Rises and floods over top of slab:
< 5 times/year   2
6 - 10 times/year   3
11 - 20 times/year   4
> 20 times/year   5

Is there any surface water <u>drain or ditch</u>
to take away rainfall?

8

- No   0
- Yes   1

32. Is the present pit replacing a full one on
<u>the same plot?</u>

9

- No   0
- Yes   1

33. <u>Is there a space on the plot to dig
another one,</u> if present pit is filled?

10

- No   0
- Yes   1

34. <u>How often is the pit desludged (emptied)?</u>

11

- Never (not yet filled)   0
- Once in every: 6 months   1
1 year   2
1.5 years   3
2 years   4
3 years   5
4 years   6
5 years and more   7

<u>Why never desludged,</u> even if filled?

12

- Lack of money   0
- Inaccessibility to desludging equipment   1
- Other reasons (.....................)   2

35. <u>Who does the desludging?</u>

13

- DKI Cleansing Department   0
- Private company (.....................)   1
- House owner   2
- Other (.................................)   3

36. <u>Where is the sludge disposed of,</u> if homeowner is doing desludging?

14
☐

- On open land   0
- Into river   1
- In other place   2

37. <u>What is the desludging fee paid</u> effectively by the family?

15
☐

    < Rp 1,000/ cu. m. of sludge   0
Rp 1,001 - Rp 2,000   1
Rp 2,001 - Rp 3,000   2
    > Rp 3,000   3

38. <u>Family's complaints</u> about leaching pit system

16
☐

- No complaints   0
- Small <u>capacity</u> of pit   1
- Desludging <u>fees</u> too high (Rp ....../cu. m.)   2
- Both of them (capacity and fee)   3
- Desludging services inadequate
  (Details ....
  ....)   4
- Other complaints (Details ....
  ....)   5

39. If another leaching pit is needed, how much would the family be ready and able to <u>pay for the construction?</u>

17
☐

- Nothing   0
- Lump sum, maximum: Rp   5,000   1
        Rp   5,001 - Rp 20,000   2
        Rp 20,001 - Rp 40,000   3
        Rp 40,001 - Rp 60,000   4
- Monthly rates, maximum: Rp   100   5
        Rp 101 - Rp   200   6
        Rp 201 - Rp   500   7
        Rp 501 - Rp 1,000   8
- Otherwise (Details ....
  ....)   9

40. If desludging fee is too high, how much would the family be ready and able <u>to pay</u> <u>for it?</u>

18
☐

- Nothing   0
- Maximum Rp   500 cu. m. of sludge   1
       Rp   750   2
       Rp 1,000   3
       Rp 1,500   4
       Rp 2,000   5

SOLID WASTE

19
☐

41. What is done with <u>rubbish</u>?

   a) Collected by <u>DKI Cleansing Department</u>
      (Dinas Kebersihan), whereby

      - rubbish is picked up from the house by hand carts   0

      - delivered by family to a storage/transfer point for
        truck collection   1

   b) Collected by <u>local community system</u>, run by the R.T's
      (Rukun Tetanggas) or R.W's (Rukun Wargas), whereby

      - rubbish picked up from the house by hand carts   2

      - delivered by family to a storage/transfer point for
        truck collection   3

   c) Thrown by family on the roadside, open land   4

   d) Thrown into drains, open ditches   5

   e) Thrown into river   6

   f) Thrown into open pit   7

   g) Otherwise (Details ....)   8

42. What are the <u>solid waste facilities</u> used by the house owner?

20
☐

   - Concrete bins   0
   - Oil drums mounted on a stand   1
   - Paper bags   2
   - Garbage cans   3
   - Others (Details ....)   4

21
☐

43. If rubbish collected, what <u>fee</u> is paid for it?

         ❮ Rp 1,000/per month   0
  Rp 1,000 - Rp 1,500   1
  Rp 1,501 - Rp 2,000   2
  Rp 2,001 - Rp 2,500   3
  Rp 2,501 - Rp 3,000   4
         ❯ Rp 3,000   5

44. If rubbish not collected, how much would the family be
    <u>willing and able to pay</u> for the rubbish collection

22
☐

   - Nothing   0
   - Maximum: Rp   500/per month   1
           Rp   750   2
           Rp 1,000   3
           Rp 1,500   4
           Rp 2,000   5

EXPLANATORY NOTES

1. The reference number (for example Ref. No. 3/8) serves for classification of
    questionnaires; later, it facilitates the search for the collected informa-
    tion for detailed planning purposes. In the questionnaire used, the first
    number refers to the survey group (No. 3), the second one is the serial
    number of households interviewed by the group in the survey zone (No. 8).
    Any other reference system which serves the above purposes can, of course,
    be used.

2. Answers and punch cards: The possible answers are coded, in some few cases
    only (water consumption, number of rooms, occupants and other items) the
    answer is to be given in absolute figures. If the answer's code or expected
    absolute figure is only one figure (from 0 to 9), one column of the punch
    80-column card has to be reserved for the answer. Similarly, if the possible
    answers require two (10 to 99), three (100 to 999) or more figures, the
    number of the punch card columns must be reserved accordingly. After all 80
    columns of the punch card have been used, other card(s) can be introduced
    and used in the same way. For punching and control purposes, it is desirable
    to designate each punch card in its first column by the appropriate number.

3. House/Plot sketch: No detailed engineering drawing is required; a hand out-
    lined approximate sketch (layout) of the house on the plot will be enough
    for illustrating the location of house and its sanitation facilities. If
    some facility is not covered by preprinted symbols, additional symbols could
    be used by surveyors.

4. Distance of house to vehicular road (Item 2): the shortest distance through
    the street(s), not a bee-line or as the crow flies, these data are relevant
    for desludging services (the length of desludging trucks' hoses are mostly
    about 80 meters).

5. Number of rooms (Item 4): all kinds of rooms(living/sitting room, bedroom,
    dining room, kitchen), excluding bathroom, closet and small auxiliary
    rooms.

6. Rent for room (Item 11): Rent paid for the house, divided by the number of
    rooms. This information can, however, be omitted, if the family is reluctant
    to give it. This could serve only as supporting data for the family income
    estimate.

7. Family income (Item 12): Income of all family members from all resources. If
    the family is hesitant to give this information, the surveyors can do their
    own estimate on the basis of their observations (general living standard of
    the family, family expenditures for house/room, water supply, etc.) and
    their own experience. This estimate can even be done at the end of the
    interview.

8. <u>Pressure of Water</u> (Item 15): If no water pressure gauge is available, the possible answers could be formulated (instead of bar) as follows: low-adequate-high.

9. <u>Water Consumption</u> (Item 19). Data given in other units (cubic meters, tins, etc.) should first be converted into liters. If the family is not in a position to give this information, the surveyors can do their own estimates.

10. <u>Delsudging aspects</u> (Items 28/c to 28/g and 34 to 40) could be dealt with under a separate chapter for both septic tanks and leaching pits together, if no special attention has to be paid to a particular one of these facilities.

11. <u>Desludging fees</u> (Items 28/g & 37), including any extra payments for services required by the desludging crew (over official rates).

SURVEY SANITASI - JSSP

A - Fasilitas Sanitasi Rumah

*) GROUP - 2
**) CASE - 2

Petugas : *Edward Malau & Muhardi* Tanggal *4-3-1982*

Kecamatan : *Setiabudi* Kelurahan *Karet*

Kampung : *6.-Karet Pedurenan* Jalan *A.11* No.Rumah *12*

Pemilik : *Atma*

Jumlah penghuni yang diwawancarai : *Atma*

Denah Rumah - Sket   (Skala: 1 : ....)

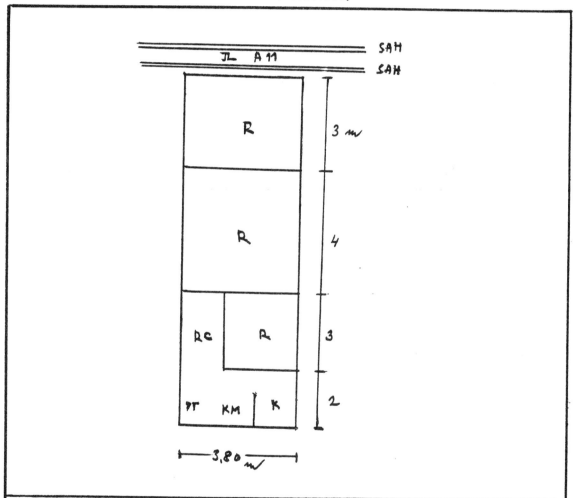

R = Ruangan	PT = Pompa Tangan
D = Dapur	SG = Sumur Gali
KM = Kamar Mandi	PT = Pipa Tegak (di plat)
RC = Ruang Cuci	⦿ = Sambungan PAM (jika ada)
K = Kakus	TA = Tangki Air
BR = Bidang Resapan	SAH= Saluran Air Hujan
ST = Septic Tank	S = Sampah
=	=

1. <u>ZONE</u> (K = Kampung; NK = Non-Kampung)

CARD-1 1 Zone |1| |:3|

K -	1	Kampung	1	Dukuh Setiabudi	1
	2		4	Karet Belakang	2
	3		6	Karet Pedurenan	3
	4		7	Kuningan I	4
	5		9	Kuningan II	5
	6		10	Kuningan III	6
	7		11	Karet Sawah	7
	8		13	Kawi Gembira	8
	9		14	Menteng Wadas I	9
	10		16	Menteng Atas	10
	11		17	Menteng Rawa Panjang	11
	12		18	Kebon Obat	12
	13		21S	Warung Pedok	13
	14		22	Warung Pedok II	14
	15		24W	Manggarai Barat/Timur	15
	16		25	Bali Matraman	16
	17		27	Bukit Duri Selatan	17
	18		28E	Melayu Kecil/Bukit Duri	18
	19		32	Tebet Timur	19
	20		36	Kebon Baru	20
NK -	1	Daerah Non Kampung	9-A		21
	2		12-A		22
	3		22-B		23
	4		25-A		24
	5		38-B		25

RUMAH DAN KELUARGA

2. <u>Jalan Keluar:</u> - Jalan Raya   0
              - Gang dengan pemadatan   1
              - Gang tanpa pemadatan   2

4 |1|

   Bila Gang, <u>berapa jarak dari rumah ke Jalan Raya</u> ?

   < 25 m   0
   25 - 50 m   1
   51 - 75 m   2
   76 -100 m   3
   >100 m   4

5 |0|

3. <u>Keadaan Rumah</u>

   - Permanent (pasangan bata, konstruksi beton)   0
   - Semi permanent (pondasi keras, bersifat sementara) 1
   - Temporary (konstruksi bambu atau kayu)   2
   - Transient (bangunan sementara, ukuran kecil)   3

6 |2|

4. <u>Jumlah ruangan</u> yang ada didalam rumah

7 | |4|

5. <u>Jumlah kepala keluarga</u> dalam rumah

8 |1|

6. Jumlah penghuni didalam rumah — — — — — — — — — — — — — — — | **4** |

7. Jumlah penghuni dewasa (umur lebih dari 15 tahun) — — — — — — — | **3** |

8. Jumlah anak2 umur dibawah 5 tahun — — — — — — — — — — — | : |

                antara 5 - 15 tahun — — — — — — — — | :**1** |

9. Pekerjaan Kepala Keluarga :

    | **3** |

- Dagang, Niaga   0
- Pengrajin (.............................)   1
- Bertani, Nelayan   2
- Buruh   3
- Bidang Administrasi, Guru   4
- Militer, Polisi   5
- Lainnya (................................)   6
- Tanpa pekerjaan   7

10. Apakah penghuni adalah <u>pemilik rumahnya sendiri</u> ?   Tidak   0 | **1** |

                                                        Ya   1

11. Bila ruangan2 disewakan, <u>berapa sewa perkamar</u> perbulannya ? | |

	< Rp. 10.000	0
10.001	Rp. 20.000	1
20.001	Rp. 40.000	2
40.001	Rp. 60.000	3
	> Rp. 60.000	4

12. Perkiraan(ancer2) <u>pendapatan keluarga</u> per bulan : | **0** |

	< Rp 30.000	0
Rp 30.000	- Rp. 55.000	1
Rp 55.001	Rp 120.000	2
Rp 120.001	Rp 200.000	3
	> Rp 200.000	4

<u>AIR BERSIH</u>

13. Bagaimana cara <u>mendapatkan air bersih</u> ?

    a)   untuk <u>minum/memasak</u> | **8** |

- Sambungan air kerumah (PAM)   0
- Satu pipa tegak (di plot)   1
- Sambungan langsung/tangki air: Umum (KIP)   2
                                                 Kongsi   3
- Sumur Dalam (> 8 m):   Umum (KIP)   4
                           Kongsi/kelompok   5
                           Pribadi   6
- Pompa Tangan, sumur dangkal (< 8 m): Umum   7
                                         Pribadi   8
                                         Kongsi   9
- Sumur gali, memakai timba : Pribadi   10
                           Kongsi/Kelompok   11
- Membeli eceran   12
- Sumber lain (................................)   13

ANNEX 2

JAKARTA SEWERAGE AND SANITATION PROJECT (JSSP)
SANITATION SURVEY

QUESTIONNAIRE B - PUBLIC WATER TAPS

A) INTRODUCTION

B) QUESTIONNAIRE

C) EXPLANATORY NOTES

D) FILLED-IN QUESTIONNAIRE IN LOCAL LANGUAGE
   (FIRST 2 PAGES)

INTRODUCTION

The main objectives of the investigation of public water taps are to determine their physical conditions, their operation and maintenance systems and if they are serving the poor population in accordance with planning intentions and, consequently, what improvements should be introduced in order to meet adequately the water demand within the project area.

Before starting any field work, it is necessary to collect from the appropriate water supply agency all relevant data related to public water taps within the area, such as their location, type and capacity, sources of water supply (i.e. city mains, autonomous deep wells ...), location and types of distribution network, registered concessionaires, managers and/or operators of taps, water charges for concessionaires, direct beneficiaries and vendors, data about water sold and other aspects. These data facilitate considerably finding the taps, identifying the persons for interview and checking if the operation of the tap is in conformity with the concession agreement.

The most appropriate persons to be interviewed will be the concessionaire, manager and/or operator of the tap, as well as some vendors and/or direct users of water. If the tap is not operating (for any reason) or operating, but without any concessionaire, manager or operator, the information could be collected from neighbors, local authorities and possible or actual beneficiaries.

After the field work is finished, the relevant survey findings (e.g. no operation of the tap, low pressure of water, needs for repairs, extension of distribution networks, establishment of new taps, and modification of water charges, etc.) are to be discussed with the agency concerned, in order that the proposed improvements will be in line with the overall development program of the municipality water supply scheme (development of water sources, extension of city mains, etc.)

Ref. No.: .../...

JSSP - SANITATION SURVEY

A - Public Water Taps

Surveyor(s): .... Date ....

Kecamatan: .... Kelurahan ....

Kampung: .... Zone ....

Tap's location (show also on the map): Street:...............................

Person(s) interviewed: ....

Card - 1

1 Zone

1. Who established the tap?

    - PAM (Municipality Water Supply Company)   0
    - Under KIP   1
    - Other (...............................)   2

2. What were the planning/design criteria for
    establishing the tap?

    - Population density (persons per ha ....)   0
    - Production capacity of water (cu. m./day ....)   1
    - Other ....)   2

3. Age of the tap (established in year ....)

        < 2 years   0
        2 - 4 years   1
        5 - 6 years   2
        7 - 10 years   3
       11 - 15 years   4
        > 15 years   5

    The tap is in operation (start in ....)

        < 2 years   0
        2 - 4 years   1
        5 - 6 years   2
        7 - 10 years   3
       11 - 15 years   4
        > 15 years   5
    No operation (Why: ....)   6

4. How many families take water directly from the tap?

    How many vendors take water for selling?

5. How much water (in cu. m./day) is taken from the tap?

    How much of this water is used for:

    - drinking/cooking (rough estimate in % ....)
    - other purposes (in % ....)

6. Water charge paid by the owner of the tap: (Rp ..../cu. m.)

7. What is the price of water sold at the tap:

    a) For people coming directly to the tap (Rp ...../2 tins)

```
        < Rp 20/2 tins (18 ltr = 1 tin)     0
  Rp 20 - Rp 40                              1
  Rp 41 - Rp 60                              2
        > Rp 60                              3
  No payment                                 4
```

    b) For vendors (Rp .... /2 tins)

```
        < Rp 20/2 tins                       0
  Rp 21 - Rp 40                              1
  Rp 41 - Rp 60                              2
        > Rp 60                              3
```

8. What is the price of water sold by vendors: (Rp ....../2 tins)

```
        < Rp 20/2 tins                       0
  Rp 41 - Rp 60                              1
  Rp 61 - Rp 70                              2
        > Rp 70                              3
```

9. How are the prices set for water at the tap?

```
  - By PAM                                   0
  - By DKI                                   1
  - By owner/concessionaire of the tap       2
  - Otherwise (........................)      3
```

10. Who is the tap manager?

```
  - Person selected by PAM                   0
  - Other (selected by ....................) 1
```

11. Who is actually operating the tap?

```
  - Tap manager himself                      0
  - Person selected by the tap manager from the
    house close to the tap                   1
  - Other person selected (by ................) 2
```

    Hours of operation (from .... to ......)

    Total number of hours of operation:...............

12. What is <u>the pressure of water?</u>

26

No pressure (no water)	0
< 0.5 bar	1
0.5 - 1.0 bar	2
1.1 - 1.5 bar	3
1.6 - 2.0 bar	4
> 2.0 bar	5

13. How often is the <u>low pressure</u> (<0.5 bar):

27

- Permanently	0
- Sporadically (........... times/day)	1

14. If inadequate pressure, is there <u>storage basin</u> at the tap?

28

- No	0
- Yes, with capacity:	
< 1 cu. m.	1
1 - 2 cu. m.	2
> 2 cu. m.	3

<u>When</u> is the storage basin filled (time: ......)
and with <u>how much water</u> (about: .... cu. m.)

15. Who are the <u>vendors?</u>

29

- Relatives of the tap manager/operator	0
- Persons without any relationship to the tap manager/operator	1
- Others selected by	2

16. <u>Does the RT/RW himself exercise surveillance</u> over the <u>tap management?</u>

30

- Yes	0
- No, no management control exists	1
the control is made by: PAM	2
DKI	3
...	4

17. <u>Physical conditions</u> of tap (estimate of surveyors):

31

- Very good	0
- Good	1
- Poor, while tap is very old	2
maintenance neglected	3
not in operation	4
other reason (...............)	5

18. Is there <u>a competition</u> in the tap "business"
    (public/private)?

  - No   0
  - Yes (.....................................) 1

19. Is the tap <u>location appropriate?</u>

  - Yes   0
  - No (why ....) 1

  If not, where should it be located (show also in
  the map):
  ....

20. Is <u>the tap used primarily</u> for serving <u>the poor</u>
    <u>families or as a "private buisness"</u> for selling
    water to customers (i.e. for generating money)
    beyond the planning intention?

  - For serving poor families   0
  - For "private business"   1

21. Is the tap <u>licensed by PAM?</u>

  - Yes   0
  - No (Why ....) 1

22. Does all water pass <u>through the meter?</u>

  - Yes   0
  - No   1

  Give estimates of water passed through meter (cu. m./day)

  and bypassing the meter (cu. m./day)

23. Have <u>PAM officials</u> ever controlled the physical conditions
    and functioning of the tap?

  - Not yet   0
  - Yes, when: (.................................) 1

24. Who furnished <u>the land</u> for the tap?

  - Private owner   0
  - Municipality   1
  - Other (.................................) 2

  Did this influence the selection of the tap
  holder or the tap manager?

  - No   0
  - Yes   1

32 □
33 □
34 □
35 □
36 □
37 □□
39 □□
41 □
42 □
43 □

25. How <u>many families</u> are living <u>in the service area</u>?

   Total number of families ....

   thereof: Poor families ....

         Families with house connection ....

26. Is the tap furnishing water to all <u>poor families</u> in the service area

      - Yes   0
      - No, because the tap is too far   1
                  the water charge is too high   2
                  the tap is out of operation   3

   If not, how do poor families in the service area
   get their drinking and cooking water?

      - From shallow wells with hand pumps and/or dug wells 0
      - Otherwise (...................................)   1

27. How many families in the service area use <u>shallow wells</u>
   (both with hand pumps and dug wells)

28. <u>Who built these wells?</u>

      - Under KIP   0
      - With self-help   1
      - Private contractor   2
      - Other   3

29. Generally, how is the <u>physical status of these wells</u>:

      - Very good (well maintained)   0
      - Good   1
      - Poor, inadequate   2

30. <u>Preliminary findings</u>

      a) Additional water taps needed (show location also in   No   0
         the map)   Yes   1

      b) If new taps needed, is the land available?   No   0
                                                                 Yes   1

      c) O&M to be improved   No   0
                                                                 Yes   1

      d) Water rates policy to be changed   No   0
                                                                 Yes   1

      e) Control to be intensified   No   0
                                                                 Yes   1

      f) Water supply (pressure) to be improved   No   0
                                                                 Yes   1

      g) Repairing needs (.................)   No   0
                                                                 Yes   1

EXPLANATORY NOTES

1. Tap's Location: This must be shown on the map (extracts of the map or sketches handed over to surveyors) on the basis of information received from the appropriate water supply agency. However, if this preliminary information does not correspond to the reality, the correction of tap's location should be notified by the surveyors. This should also be done and indicated in the map/sketch, if the actual location of the tap has to be modified (Item 19) or if new taps are proposed (Item 30).

2. Quantity of water taken from the tap (Item 5): If not registered by operator/concessionaire, a rough estimate (an average in cu. m./day) can be given by operator/concessionaire or calculated by surveyors.

3. Pressure of water (Item 12): If no water pressure gauge is available, the possible answers could be formulated (instead of bar) as follows: low - adequate - high.

4. Families within the service area and level of water supply services (Items 25 to 29): The required data and information could be obtained (at least as estimates) from the appropriate local authorities.

5. Preliminary findings (Item 30): Preliminary judgements of surveyors which could help the planning of improvements (but not indispensable).

JSSP - SANITATION SURVEY
B - Keran Umum (K.U.)

Petugas : *RUSYDI RUSLI + SUHENDI* .... Tanggal *18-3-1982* ....
Kecamatan : *TEBET* .... Kelurahan : *MENTENG DALAM* ....
Kampung *22 - WARUNG PEDONG* ....Zone ....*14* ....
Lokasi K.U.(Perlihatkan pada peta); Jalan : ....*B* ....
Jumlah penduduk yang diwawancarai : ...*4 orang., RT* ....

CARD-1

1 zone		
1	1	4

1. Siapa yang membangun K.U.?

   - PAM   0
   - KIP   1
   - Lainnya (.........................)   2

4
1

2. Apa saja kriteria perencanaan untuk membangun K.U.?

   - Kepadatan penduduk   0
   - Kapasitas air yang diproduksi (m3/hari ......)   1
   - Lainnya : *KHUSUS UNTUK MUSOLAH*   2

5
2

3. Umur K.U. (dibangun pada tahun 19...*1976*)

< 2 tahun	0
2 - 4 tahun	1
5 - 6 tahun	2
7 - 10 tahun	3
11 - 15 tahun	4
> 15 tahun	5

6
2

   Keran Umum sudah bekerja (mulai tahun ......*1976*.)

< 2 tahun	0
2 - 4 tahun	1
5 - 6 tahun	2
7 - 10 tahun	3
11 - 15 tahun	4
> 15 tahun	5
Tidak bekerja	6

7
2

4. Berapa jumlah keluarga yang mengambila air langsung dari K.U.?

9		
	1	0

   Berapa jumlah pengecer air yang mengambil air dari K.U. ?

11	

5. Berapa jumlah air (dalam m3/hari) yang diambil dari K.U. ?

13	
	4

   Berapa bagian dari seluruh jumlah air yang diambil dari K.U.
   dipergunakan untuk :

   - minum/masak (perkiraan kasar : ....%) _ _ _ _ _ _ _ _

15	
	1

   - keperluan lain (..............%) _ _ _ _ _ _ _ _ .

17	
	3

6. Harga air yang ditetapkan oleh PAM untuk K.U.:(Rp........./m3)

7. Berapa harga air yang dijual di K.U. ?

    a) untuk orang yang datang langsung :(Rp....../pikul)

< Rp 20/pikul (1 pikul = 36 ltr)	0
Rp 20 - Rp 40	1
Rp 41 - Rp 60	2
Rp 60	3
Tidak bayar	4

$\underset{\boxed{4}}{\overset{19}{}}$

    b) untuk pengecer air : (Rp.../pikul)

< Rp 20/pikul (2 x 16 liter)	0
Rp 21 - Rp 40	1
Rp 41 - Rp 60	2
> Rp 60	3

$\underset{\boxed{}}{\overset{20}{}}$

8. Berapa harga dijual oleh pengecer air ?

< Rp 40/pikul (2 x 16 ltr)	0
Rp 41 - Rp 60	1
Rp 61 - Rp 70	2
> Rp 70	3

$\underset{\boxed{}}{\overset{21}{}}$

9. Siapa yang menetapkan harga air di K.U. ?

- oleh PAM	0
- oleh PEMDA DKI	1
- oleh pemilik K.U.	2
- Badan lainnya (.........................)	3

$\underset{\boxed{}}{\overset{22}{}}$

10. Siapakah pengelola (manager) dari K.U. ?

- Seseorang yang dipilih oleh PAM	0
- Lainnya (dipilih oleh *PENGURUS. MUSOLAH*)	1

$\underset{\boxed{1}}{\overset{23}{}}$

11. Siapa sebenarnya yang mengoperasikan/mengelola K.U. ?

- Manager (Pengelola) K.U. sendiri	0
- Seseorang yang dipilih oleh Manager K.U. dan orang ini bertempat tinggal dekat K.U.	1
- Orang lain (dipilih oleh ....)	2

$\underset{\boxed{1}}{\overset{24}{}}$

    Lamanya bekerja (K.U.) : dari jam **4.00** sampai jam **21.00**

    Jumlah jam kerja/hari: _ _ _ _ _ _ _ _ _ _ _ _ _ _ _ _ _ _ _

$\underset{\boxed{1\,7}}{\overset{25}{}}$

12. Berapa tekanan air yang ada pada K.U. ?

- Tidak ada tekanan sama sekali (air tidak mengalir)	0
< 0,5 bar	1
0,5 - 1,0 bar	2
1,0 - 1,5 bar	3
1,6 - 2,0 bar	4
> 2,0 bar	5

$\underset{\boxed{1}}{\overset{27}{}}$

ANNEX 3

JAKARTA SEWERAGE AND SANITATION PROJECT (JSSP)

SANITATION SURVEY

QUESTIONNAIRE C - MCK (COMMUNAL SANITATION FACILITIES)

A) INTRODUCTION

B) QUESTIONNAIRE

C) EXPLANATORY NOTES

D) FILLED-IN QUESTIONNAIRE IN LOCAL LANGUAGE
(FIRST 2 PAGES)

INTRODUCTION

Similarly to public water taps, the main aim of surveying communal sanitation facilities (toilets, bathing, washing) is to determine their physical conditions, their operation and maintenance systems and their acceptance by the community and, consequently, what improvements should be introduced in order that they effectively serve the poor population in accordance with planning intentions.

Before starting this survey, it is necessary to collect from the appropriate municipal department and/or local authorities all relevant data related to these facilities, such as their location, types and capacities, those responsible for their management, operation and maintenance, and other aspects. These data help finding the facilities, identifying the persons for interviews and checking if the facilities are used as initially intended.

The appropriate persons to be interviewed will be the assigned manager and/or operator of the facility, as well as direct beneficiaries. If the facility is not in operation (for any reason), the information could be collected from the neighbors, local authorities and intended beneficiaries.

Preliminary survey findings (e.g., no operation of the facility, poor cleanliness, inadequate water supply, needs for repairs, modification of services charges, etc.) are to be discussed immediately, or later, with the appropriate authorities and people's representatives in order that the proposed improvements will be accepted by the people concerned and in line with the overall sanitation program of the municipality.

JSSP - SANITATION SURVEY

C - MCK (Communal Sanitation Facilities)

Surveyor(s): .... Date: ....

Kecamatan: .... Kelurahan: ....

Kampung: .... Zone No.: ....

MCK's location (show also on the map), Street: ....

Person(s) interviewed: ....

MCK - Sketch   Card 1

1 Zone

1		

Front & Side Elevation	
Ground Plan	

WT = Water Tap	T = Toilet Room
DW = Deep Well	B = Bathroom
HW = Handpump Well	W = Washing Unit
WZ = Water Tank	=

1. <u>Who has established the MCK?</u>

    4
    ☐

- PAM   0
- Public Works   1
- Health Department   2
- DKI Cleansing Department   3
- Under KIP   4
- Other (...............................)   5

2. What were the <u>planning/design criteria</u> for establishing MCK
(population = poor population without house sanitation facilities)

    5
    ☐

- One MCK per: ❮ 1,000 population   0
            1,001 - 2,500   1
            2,501 - 5,000   2
            5,001 - 10,000   3
        10,001 - 15,000   4
              ❯ 15,000   5

3. What is <u>the actual "service area"</u> of MCK
(............... population)

    6
    ☐

             ❮ 2,000 population   0
        2,001 - 5,000   1
        5,001 - 10,000   2
     10,001 - 15,000   3
         ❯ 15,000   4

4. <u>Age of MCK</u> (established in year ....)

    7
    ☐

    ❮ 2 years   0
  2 - 4 years   1
  5 - 10 years   2
11 - 15 years   3
  ❯ 15 years   4

MCK is on <u>operation</u> (start in ....)

    8
    ☐

    ❮ 2 years   0
  2 - 4 years   1
  5 - 10 years   2
11 - 15 years   3
  ❯ 15 years   4
    No operation (why ....)   5

5. <u>How many families use the MCK?</u> (rough estimates)

    9

a) for water supply (...... tins/day), Number of families:

    12

b) for toilet (....... persons/day), Number of families:

    15

c) for bathing (....... persons/day), Number of families:

    18

d) for washing (....... persons/day), Number of families:

6. <u>What kind of water supply system</u> is in the MCK?

- Piped water supply (city mains)	0
- Deep well with electric pump	1
with windmill	2
- Hand pump, shallow well	3
- Other (dug well ....)	4

7. <u>How much water</u> (in cu. m./day, 1,000 l = 1 cu. m.)

    a) <u>is supplied to households</u>

    b) is <u>used in MCK</u> - for toilets

                    - for bathing

                    - for washing

8. What are <u>the charges</u> for MCK services?

    a) for <u>water supply</u>:

< Rp 40/2 tins	0
Rp 41 - Rp 50	1
Rp 51 - Rp 60	2
> Rp 60	3
No charge	4

    b) for <u>toilet adult</u>:

< Rp 5	0
Rp 5 - Rp 10	1
Rp 11 - Rp 15	2
Rp 16 - Rp 20	3
> Rp 20	4
No charge	5

                <u>child</u>:

< Rp 5	0
Rp 5 - Rp 10	1
Rp 11 - Rp 15	2
> Rp 15	3
No charge	4

    c) for <u>bathing</u> - <u>adult</u>:

< Rp 10	0
Rp 10 - Rp 15	1
Rp 16 - Rp 20	2
Rp 21 - Rp 25	3
> Rp 25	4
No charge	5

34 ☐

child:

≺ Rp 5	0
Rp 5 - Rp 10	1
Rp 11 - Rp 15	2
≻ Rp 15	3
No charge	4

35 ☐

d) for <u>washing</u> (Rp/.........)

≺ Rp 20	0
Rp 21 - Rp 25	1
Rp 26 - Rp 30	2
≻ Rp 30	3
No charge	4

36 ☐

9. <u>How are the above charges set up?</u>

- by PAM	0
- by DKI (local authorities)	1
- by Health Department	2
- Other (.....................)	3

37 ☐

10. <u>What is the money used for?</u>

- To cover O&M cost (salaries, electricity ...)	0
- Otherwise (...........................)	1
- Unknown	2

38 ☐

11. <u>Physical conditions of MCK</u> (surveyor's judgement)

- very good, well maintained	0
- good	1
- poor, inadequate	2

39 ☐

<u>If inadequate, why?</u>

- MCK is too old	0
- Maintenance neglected	1
- Lack of finance sources	2
- Combined (..........)	3
- Other (............)	4

40 ☐

<u>Impacts of the (poor) state of repair:</u>

- MCK out of operation	0
- MCK's operation reduced	1
- MCK's quality of services is very low	2
- Losses of customers	3
- Combined (................)	4
- Other (..................)	5

41 ☐

12. <u>State of cleanliness of MCK:</u>

- Always kept clean	0
- Sometimes uncleaned (....... times/month)	1
- Always uncleaned	2

If uncleaned (Codes 1+2), why?

- MCK's staff insufficient   0
- O&M not controlled   1
-   2

42 ☐

Impacts of the poor cleanliness of MCK:

- MCK's quality of services very low   0
- Losses of customers   1
- Combined (................)   2
- Other (..................)   3

43 ☐

13. Status of the MCK Water Supply:

- Very good, well maintained   0
- Good   1
- Poor, inadequate   2
- No water supply   3

44 ☐

If poor status of repair, why?

- Water supply system very old   0
- Maintenance neglected   1
- Lack of financial sources   2
- Combined (..............)   3
- Other (................)   4

45 ☐

14. What is the pressure of water, if piped water supplied:

No pressure (no water)   0
   < 0.5 bar   1
0.5 - 1.0 bar   2
1.1 - 1.5 bar   3
1.6 - 2.0 bar   4
  > 2.0 bar   5

46 ☐

If inadequate pressure, is there any storage basin?

- No   0
- Yes, with capacity:
  < 1 cu. m.   1
 1 - 2 cu. m.   2
  > 2 cu. m.   3

47 ☐

When is the storage basin filled (time: ....)
and with how much water (about .... cu. m.)

15. Who is the manager of the MCK:

- Person selected by PAM   0
- Person selected by RT/RW   1
- Other (selected by ....) 2
  What kind of work does he do?
  (describe: ....)

48 ☐

49 ☐

16. Who is actually operating the MCK:

    - The MCK Manager himself   0
    - Person selected by the MCK Manager   1
    - Other person selected by   2

    Time of operation: (from .... to ....)
    Total number of hours in operation per day ....

50 ☐☐

17. Did the PAM Officials ever control the physical conditions and function of the water supply?

    - Never   0
    - Yes, regularly (............ times/year)   1
             irregularly (..................)   2

52 ☐

18. Does the RT/RW himself exercise surveillance over the MCK management/operation?

    - Yes   0
    - No   1

53 ☐

19. What kind of toilet system is in the MCK?

    - Water seal with pour flush   0
              with cistern flush   1
    - Squatting plate, with pour flush   2
                  with cistern flush   3
    - Other (.......................................)   4

54 ☐

    Is the liquid effluent from the toilets going:

    - to sewer   0
    - to septic tank with drain fields   1
    - to septic tank with overflow to drains/ streams   2
    - direct discharge to drains/streams/rivers   3
    - other (................................)   4

55 ☐

20. Number of toilet units:

    2 units   0
    4   1
    6   2
    8   3
    10   4
    12   5
    16   6

56 ☐

21. Number of laundry units:

    2 chambers   0
    4   1
    6   2
    8   3

57 ☐

58

22. <u>Number of bathing units:</u>

    2 units   0
    4   1
    6   2
    8   3
    10   4
    12   5

59

23. <u>Number of employees:</u>

    1   0
    2   1
    3   2
    4   3
    No employee (self-maintenance by users)   4

60

24. <u>Who are the customers of the MCK?</u>

    - Transients   0
    - Residents of the service area   1
    - Both   2

25. Total <u>number of poor families</u> in the MCK
    service area, who do not have house toilet/
    washing/bathing facilities (estimates):

    61

    Thereof: families using MCK (estimate)

    64

    Where do the others defecate? Mostly:

    67

    - Friends (shared) latrine   0
    - Unoccupied land   1
    - River   2
    - Drains, ditches, canals   3
    - Other (...................)   4

    68

    <u>Where do the others bathe/wash?</u> Mostly:

    - Friends (shared) facilities   0
    - River   1
    - Other (...................)   2

69

26. The families using the MCK in the service
    area, <u>how far do they have to walk?</u> Mostly:

         < 25 m   0
    26 - 50 m   1
    51 - 100 m   2
    101 - 200 m   3
        > 200 m   4

70

27. What is <u>the maximum distance</u> people are
    willing to walk to use the MCK?

        50 m   0
      100 m   1
      150 m   2
      200 m   3
    > 200 m   4

28. Who furnished the land for the MCK?

    - Private owner   0
    - Municipality   1
    - Other (..................)   2

    Did this influence the <u>location and size of the MCK</u>?

    - No   0
    - Yes   1

    Did this influence the <u>selection of the MCK</u>
    <u>manager/operator/holder</u>?

    - No   0
    - Yes   1

29. <u>Did the MCK meet its intended objective?</u>

    - Yes   0
    - No, because of small size   1
                bad location (long distance) 2
                lack of privacy   3
                O&M neglected   4
                combined (code: ....) 5
                other reasons (...........) 6

30. <u>Preliminary findings:</u>

    a) Additional MCKs needed (notify number of   No 0
       toilet units and show location on the   Yes 1
       map)

    b) If new MCK needed, is the land avail-   No 0
       able?   Yes 1

    c) O&M to be improved   No 0
                                     Yes 1

    d) Service rates policy to be changed   No 0
                                     Yes 1

    e) Control to be intensified   No 0
                                     Yes 1

    f) Water supply to be improved   No 0
                                     Yes 1

    g) Repairing needs (...................   No 0
       ....)   Yes 1

    h) Design modification needed   No 0
       (.................................   Yes 1
       ....)

    i) Other (.........................)

CARD-2

86

EXPLANATORY NOTES

1. <u>MCK-Sketch:</u> No detailed engineering drawing is required; a hand outlined approximate sketch (layout) of the MCK will be enough for illustrating the type of facility.

2. <u>Quantity of water supplied/used (Item 7):</u> If not registered by the operator, rough estimates (an average in cu.m./day) can be given by him or calculated by surveyors.

3. <u>Pressure of water (Item 14):</u> If no water pressure gauge is available, the possible answers could be formulated (instead of bar) as follows: low - adequate - high.

4. <u>Families within the service area, land for MCK</u> and related questions (Items 25 to 29): The required data and information could be obtained (at least as estimates) from the appropriate local authorities.

5. <u>Preliminary findings (Item 30):</u> Preliminary judgements of surveyors which could help the planning of improvements (but not indispensible).

JSSP - SANITATION SURVEY
C - MCK

Petugas : *ABDUL SYUKUR/ TJUT NASTRI H.* ..... Tanggal: **23.3.1982**
Kecamatan : *SETIA BUDI* ..... Kelurahan : *MENTENG ATAS*
Kampung *17.MENTENG RAWA PANJANG* ..... Zone No.: *11*
Lokasi MCK (tunjukkan pada peta), Jalan : *RELA 5*
Jumlah penduduk yang diwawancarai : *7 ORANG — RT*

Gambar MCK (Sket)   *CARD 1*   **1 Zone** | 1 | 1 | 1 |

Sb	= Sambungan	K	= Kakus
SDl	= Sumur Dangkal	KM	= Kamar Mandi
SPT	= Sumur Pompa Tangan	KC	= Kamar Cuci
TA	= Tangki Air		=

1. Siapakah yang membangun MCK ?

    - PAM   0
    - P.U.   1
    - Dinas Kesehatan   ·   2
    - Dinas Kebersihan DKI   3
    - KIP   4.
    - Lainnya (..........................)   5

2. Apa saja kriteria perencanaan untuk pengadaan sebuah MCK
   (penduduk = penduduk tidak mampu, tanpa fasilitas sanitasi)

    - Sebuah MCK per :   < 1.000 penduduk   0
                     1.001 - 2.000 penduduk   1
                     2.001 - 3.000 penduduk   2
                     3.001 - 5.000 penduduk   3
                     5.001 - 10.000 penduduk   4
                    10.001 - 15.000 penduduk   5
                    15.001 - 20.000 penduduk   6
                         > 20.000 penduduk   7

3. Berapa besar sebenarnya daerah pelayanan sebuah MCK ?(.... penduduk)

         < 1.000 penduduk   0
    1.001 - 3.000   1
    3.001 - 5.000   2
    5.001 - 10.000   3
    10.001 - 15.000   4
    15.001 - 20.000   5
         > 20.000   6

4. Umur MCK (dibangun tahun **197.2.**)

      < 2 tahun   0
    2 - 4   tahun   1
    5 - 6   tahun   2
    7 - 10 tahun   3
    11 - 15 tahun   4
      > 15 tahun   5

   MCK telah bekerja selama:(mulai tahun : **7.9.7.2**)

      < 2   tahun   0
    2 - 4   tahun   1
    5 - 6   tahun   2
    7 - 10 tahun   3
    11 - 15 tahun   4
      > 15 tahun   5
    Tidak bekerja   6

5. Berapa jumlah keluarga yang menggunakan MCK (perkiraan kasar)

    a) sebagai penyediaan air bersih (........... kaleng/hari)

    b) untuk keperluan kakus (WC)   (..**400**..orang/hari)

    c) untuk mandi (..**200**.. orang/hari)

    d) untuk mencuci (......... orang/hari)

ANNEX 4

JAKARTA SEWERAGE AND SANITATION PROJECT (JSSP)
SANITATION SURVEY

QUESTIONNAIRE D - SURFACE DRAINS

A) INTRODUCTION

B) QUESTIONNAIRE

C) EXPLANATORY NOTES

D) FILLED-IN QUESTIONNAIRE IN LOCAL LANGUAGE
(FIRST 2 PAGES)

INTRODUCTION

Similarly to public water taps and communal sanitation facilities (MCK), the main aim of surveying surface drains is to determine their physical condition and maintenance (cleanliness), and what measures should be proposed in order to improve the overall environmental living conditions within the area.

The work will be greatly facilitated if appropriate street maps of the area to be surveyed (with indications of drain types and slopes) are available; if not, more measurements in the field will be required and, consequently, more time will have to be devoted to the survey.

The data will be obtained mostly through the observations of the surveyors themselves; only for some small amount of information (concerning historical data and maintenance systems) should the appropriate municipal department and/or local authorities be interviewed.

Also, preliminary survey findings (e.g. poor cleanliness, need for repairs, construction of new drains, etc.) are to be discussed immediately, or later, with the appropriate authorities in order that the proposed improvements will be in line with the overall sanitation program of the municipality.

JSSP - SANITATION SURVEY

D - Surface Drains

Surveyor(s): ....Date: ....

Kecamatan: .... Kelurahan: ....

Kampung: .... Street: ....

Person(s) interviewed: ....Zone No.: ....

Drain-Sketch   Card 1   1 ZONE

Type (The narrowest cross-section of the lane)	
Slope of the drain (indicate direction of sope on the map)	4 □ - good 0 - enough 1 - not enough - length (... m) 2 - backwater due to obstacle, length (... m) 3 - ponding - length (... m) 5
Cross-sections (cm) of the drain (indicate loca-tion on the map from where to where by numbers)	

Construction material	5 □
	Earth - entirely 0
	Stone - entirely 1
	Concrete - entirely 2
	Combined: Concrete (... %), stone (... %) 3
	          Concrete (... %), earth (... %) 4
	          Stone (... %), earth (... %) 5

1. <u>Physical conditions</u> of the drain:

□ 6

   - very good   0
   - good   1
   - poor, due to: . old age   2
                    . damages (length .... m)   3
                    . narrowings   4
                    . obstruction by pipes   5
                      road widening   6
                      bridge and coverings   7
                    . maintenance neglected   8
                    . other (...............)   9

2. <u>Drain system</u> in the area was:

a) <u>designed</u> by:

□ 7

   - PAM   0
   - Public works   1
   - DKI, under KIP   2
   - Private company   3
   - Other (..............)   4

b) <u>built</u> by:

□ 8

   - PAM   0
   - Public works   1
   - DKI, under KIP   2
   - Private company   3
   - Other (..............)   4

   Cost of drains (per cu. m.) Rp ..../cu. m.

3. <u>Age of the drains</u> (built in 19....):

□ 9

   < 2 years   0
 2 - 5 years   1
 6 - 10 years   2
11 - 15 years   3
16 - 20 years   4
 > 20 years   5

4. <u>O&M of drain system</u>:

□ 10

   Responsibility of: - Public Works   0
                     - DKI   1
                     - Other (................) 2

5. <u>Role and attention of RT and RW</u> to the drains:

□ 11

   - High attention, regular control   0
   - Sporadic control   1
   - No control   2

6. The ditch system in the area <u>serves for removing:</u>   □ 12

   - Storm water only   0

   - Storm water and sullage water (from kitchen/ bathing/
     washing)   1

   - Storm water and sullage water and overflow from pits   2

   - As (2) and solid wastes   3

7. <u>Is the drain cleaned?</u>   □ 13

   - Yes   0
   - Inadequately   1

8. <u>The trash is removed:</u>   □ 14

   a) Periodically (........ times per month), effectively   0
      sporadically   1
      noneffectively   2

   b) by: - DKI - Public Works   0   □ 15
          - Self-help   1
          - Other (.............)   2

   Who pays for it? ....

   How much does it cost? Rp ....

   How is the cost calculated? ....

   ....

9. How is the trash taken from the <u>drain disposed of?</u>   □ 16

   - to designated pickup places   0
   - along the vehicular roads   1
   - along the waterways   2
   - otherwise (.....................)   3

10. <u>If cleaning is inadequate,</u> how could it be improved?   □ 17

   - by reglementation about solid wastes   0
   - by regular maintenance   1
   - by both measures (0 and 1)   2
   - otherwise (.....................)   3

11. Even if the drains are kept clean, how does it function?   □ 18

   - Very well   0
   - Capacity too small   1
   - Slope too flat   2
   - No outlet point   3

12. Is the drain in <u>the right location</u>?

    - Yes   0

    - No   1

19 ☐

    If no, why? ....

    What should be done to correct location? ....

    ....

13. <u>Is there an odor problem?</u>

    - Yes, always   0

            in dry season only   1

    - No or negligible only   2

20 ☐

14. Is the drain <u>flooded</u> during rainy days?

    - Yes   0

    - No   1

21 ☐

15. Do <u>children play in the drains</u>?

    - Yes   0

    - No   1

22 ☐

16. <u>Preliminary findings:</u>

    a) Additional drains needed   No 0
                                                      Yes 1

23 ☐

    b) Planning/design of larger drain capacities   No 0
                                                      Yes 1

24 ☐

    c) Higher self-discipline of people in the area to keep   No 0
        the drains clean   Yes 1

25 ☐

    d) Strong reglementation for disposing of solid wastes   No 0
                                                      Yes 1

26 ☐

    e) Full involvement and regular control of RW/RT in   No 0
        O&M of drains   Yes 1

27 ☐

    f) Other: ....

EXPLANATORY NOTES

1. <u>Drain-Sketch:</u> No detailed engineering drawing is required; a hand outlined approximate sketch of the drain will be enough for illustrating its type.

2. <u>Drain system</u> and related questions (Items 2 to 9): Data could be obtained from the approximate municipal department and/or local authorities.

3. Suggestions for <u>cleanliness improvement</u> (Item 10): Surveyors' judgements.

4. <u>Preliminary findings</u> (Item 16): Preliminary judgements of surveyors which could help the planning of improvements (but not indispensible).

JSSP - SANITATION SURVEY
D - Saluran Air Hujan (Drainasi)

Petugas : .**EDWARD MALAU & MUHARDI**...... Tanggal : ...**9-3-1982**...

Kecamatan : .**SETIABUDI**.............. Kelurahan ...**KARET**...................

Kampung : **4-KARET BELAKANG II**... Jalan ...**D.1**.........................

Jumlah penduduk yang diwawancarai : .... Zone No..**2**...............

Gambar(Sket) Saluran   **CARD-1** **1 ZONE** [1] [2]

Jenis/Tipe (Penampang jalan paling sempit) Ukuran dalam (cm)	

Kemiringan saluran (tandai pada peta, arah kemiringan)	- baik 0 - cukup 1 - tidak cukup , panjang (..... m) 2 - aliran balik (bila ada penyempitan) (... m) 3 - genangan , panjang (..**50** m) 4	**4** [4]

Penampang saluran (pada titik awal,tengah dan akhir (beri nomor pada peta titik² pengambilan penampang)	

Bahan bangunan yang dipakai	Tanah 0 Batu 1 Beton 2 Combinasi : Beton (...... %) + Batu (...%) 3 Beton (...... %) + Tanah(...%) 4 Batu (.... %) + Tanah(...%) 5	**5** [2]

1. <u>Keadaan fisik saluran</u> :

   - baik sekali   0
   - baik   1
   - buruk disebabkan tuanya saluran :   2
                        - rusak (panjang .... m)   3
                        - penyempitan   4
                        - kerusakan karena : diganti dengan pipa 5
                                             pelebaran jalan   6
                                             jembatan + penutupan
                                                        saluran   7
                        - perawatan diabaikan —   8
                        - lainnya (..............)   9

   ₆
   [8]

2. <u>Sistem saluran air hujan</u> (drainase) dalam daerah yang ditinjau

   a) direncanakan oleh : - PAM   0
                           - P.U.   1
                           - DKI dibawah KIP   2
                           - Perusahaan Swasta   3
                           - Lainnya (.............)   4

   ₇
   [2]

   b) dibangun oleh : - PAM   0
                       - P.U.   1
                       - DKI, dibawah KIP   2
                       - Perusahaan Swasta   3
                       - Lainnya (............)   4

   ₈
   [2]

   c) Biaya saluran (per cu.m.): Rp ....)cu.m.

3. <u>Umur saluran</u> (dibangun pada tahun .197.4...)

          2 tahun   0
      2 - 5 tahun   1
      6 - 10 tahun   2
      11 - 15 tahun   3
      16 - 20 tahun   4
          20 tahun   5

   ₉
   [2]

4. <u>O & M (Operasi dan Perawatn)</u> sistem saluran air hujan :

   Tanggung jawab : - P.U   0
                    - DKI·   1
                    - Lainnya (...........)   2

   ₁₀
   [1]

5. <u>Peranan dan perhatian RT dan RW</u> terhadap saluran :

   - Perhatian yang besar   0
   - Jarang diperiksa   1
   - Tidak ada pemeriksaan   2

   ₁₁
   [2]

6. Saluran yang ada dipergunakan untuk mengalirkan :

   - air hujan saja   0
   - air hujan dan air buangan (dari dapur, kamar mandi, cuci)   1
   - air hujan dan air buangan ditambah luapan dari cubluk   2
   - seperti (2) + buangan padat/sampah   3

   ₁₂
   [3]

7. <u>Apakah saluran dibersihkan</u> ?

   - Ya   0
   - Tidak, tidak memenuhi syarat 1

   ₁₃
   [0]

ANNEX 5

JSSP - PRACTICAL TRAINING PROGRAMME (Kampung 6 + 13)

Group	DKI/ ENCONA	9 - 10 h	10 - 11 h	11 - 12 h	12 - 13 h
1. Mr. AHMAT HAYAT Mr. MOHAMMED YAMIN	ENCONA DKI	A	B	C	D
2. Mr. EDWARD MALAU Mr. MUHARDI	ENCONA DKI	A	B	C	D
3. Mr. ACHMAD MULAWARMAN Miss SARI MUSTIKA	DKI ENCONA	B	A	D	C
4. Mr. ABDUL SYUKUR Miss CUT NASTRI HAYATI	DKI ENCONA	B	A	D	C
5. Mr. ASHARI Mr. HARYANTO	DKI ENCONA	C	D	A	B
6. Mr. ALEX KANDAR Mr. FIFI KUSUMA	DKI ENCONA	C	D	A	B
7. Mr. BUDI SOFYANHADI Mr. AMIRUL AKMAL	DKI DKI	D	C	B	A
8. Mr. RUSYDI RUSLI Mr. SUHENDI	DKI DKI	D	C	B	A

A = Housing Sanitation Facilities
B = Water Tapes
C = MCK
D = Surface Drains

JL G 7
JL G-7, JL A-4, JL-D, JL H + H1 + H2
Kampung 13
according to the map.

ANNEX 6

JAKARTA SEWERAGE AND SANITATION PROJECT, SANITATION SURVEY: ORGANIZATION CHART

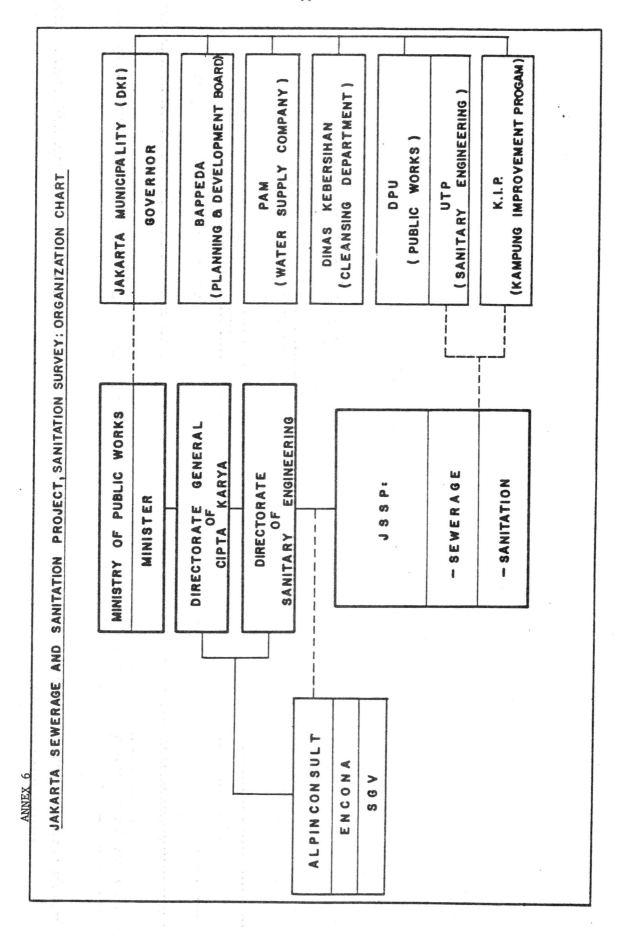

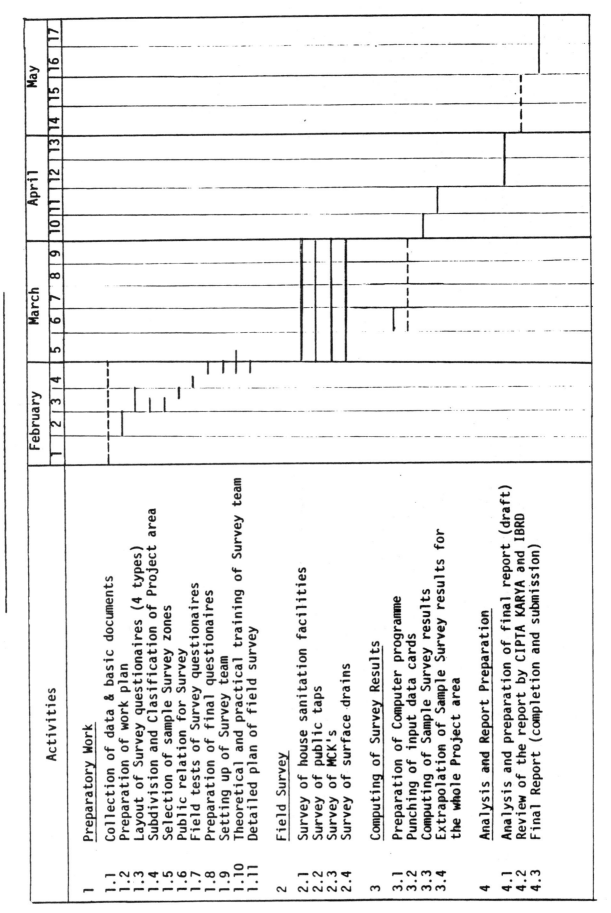

ANNEX 7
SANITATION SURVEY - TENTATIVE TIME SCHEDULE

ANNEX 8

JSSP Sanitation Survey: Computer Output for House Sanitation Facilities,
Example: Water Supply

JSSP SANITATION SURVEY - HOUSE FACILITIES
ALL KAMPUNGS
DATE: 03.31.1982

VAR.13/A WATER SUPPLY FOR DRINKING & COOKING

CATEGORY LABEL	CODE	ABSOLUTE FREQUENCY	RELATIVE FREQUENCY (PERCENT)	ADJUSTED FREQUENCY (PERCENT)	CUMULATED ADJ.FREQ. (PERCENT)
METERED HOUSE CONNEC.	0	25	1.5	1.5	1.5
STANDPIPE ON PLOT	1	38	2.3	2.3	3.8
PUBLIC WATER TAP(KIP)	2	1	0.1	0.1	3.9
SHARED WATER TAP	3	3	0.2	0.2	4.1
PUBLIC DEEP WELL(KIP)	4	28	1.7	1.7	5.8
SHARED DEEP WELL	5	86	5.1	5.1	10.9
PRIVATE DEEP WELL	6	552	32.9	32.9	43.8
PUBLIC SHALLOW WELL	7	19	1.1	1.1	44.9
PRIVATE SHALLOW WELL	8	382	22.8	22.8	67.7
SHARED SHALLOW WELL	9	47	2.8	2.8	70.5
PRIVATE DUGWELL	10	281	16.8	16.8	87.3
SHARED DUGWELL	11	167	10.0	10.0	97.3
VENDORS	12	6	0.4	0.4	97.7
OTHER SOURCES	13	41	2.3	2.3	100.0
TOTAL		1676	100.0	100.0	

VAR.13/B WATER SUPPLY FOR HYGIENIC & OTHER PURPOSES

CATEGORY LABEL	CODE	ABSOLUTE FREQUENCY	RELATIVE FREQUENCY (PERCENT)	ADJUSTED FREQUENCY (PERCENT)	CUMULATED ADJ.FREQ. (PERCENT)
AS FOR DRINK.& COOK.	0	736	43.9	43.9	43.9
PRIVATE SHALLOW WELL	1	373	22.2	22.2	66.1
SHARED SHALLOW WELL	2	52	3.1	3.1	69.2
PRIVATE DUGWELL	3	303	18.1	18.1	87.3
SHARED DUGWELL	4	165	9.9	9.9	97.2
OTHER SOURCES	6	47	2.8	2.8	100.0
TOTAL		1676	100.0	100.0	

JSSP Sanitation Survey: Computer Output for House Sanitation Facilities,
Example: Toilet Systems

JSSP SANITATION SURVEY - HOUSE FACILITIES
ALL ZONES
DATE: 04.05.1982

VAR.23 TOILET SYSTEM ON THE PLOT

CATEGORY LABEL	CODE	ABSOLUTE FREQUENCY.	RELATIVE FREQUENCY (PERCENT)	ADJUSTED FREQUENCY (PERCENT)	CUMULATED ADJ.FREQ. (PERCENT)
NONE	0	301	16.5	16.5	16.5
CISTERN FLUSH WC	1	1292	70.8	70.8	87.3
POUR SQUAT PLATE	2	170	9.3	9.3	96.6
VENTILATED LATRINE	3	15	0.8	0.8	97.4
NONVENTILATED LATRINE	4	36	2.0	2.0	99.4
OTHER	5	12	0.6	0.6	100.0
TOTAL		1826	100.0	100.0	

VAR.24/A PEOPLE EASE IF NO LATRINE ON THE PLOT

CATEGORY LABEL	CODE	ABSOLUTE FREQUENCY	RELATIVE FREQUENCY (PERCENT)	ADJUSTED FREQUENCY (PERCENT)	CUMULATED ADJ.FREQ. (PERCENT)
NEIGHBOURS LATRINE	0	64	3.5	21.3	21.3
MCK	1	73	4.0	24.2	45.5
OPEN LAND	2	3	0.2	1.0	46.5
RIVER	3	141	7.7	46.8	93.3
DRAINS,DITCHES	4	6	0.3	2.0	95.3
OTHERWISE	5	14	0.8	4.7	100.0
OUT OF RANGE		1525	83.5	MISSING	
TOTAL		1826	100.0	100.0	

VAR.26 LIQUID EFFLUENT FROM TOILET IS GOING

CATEGORY LABEL	CODE	ABSOLUTE FREQUENCY	RELATIVE FREQUENCY (PERCENT)	ADJUSTED FREQUENCY (PERCENT)	CUMULATED ADJ.FREQ. (PERCENT)
TO SEWER	0	6	0.3	0.4	0.4
TO SEPTANK & DRAIN F.	1	309	16.9	20.3	20.7
TO SEPTANK & OVERFLOW	2	180	9.9	11.8	32.5
TO LEACHING PIT	3	888	48.6	58.2	90.7
TO DRAINS DIRECTLY	4	107	5.9	7.0	97.7
TO STREAMS,RIVERS	5	30	1.6	2.0	99.7
OTHER	6	5	0.3	0.3	100.0
OUT OF RANGE		301	16.5	MISSING	
TOTAL		1826	100.0	100.0	

ANNEX 10
JSSP Sanitation Survey: Computer Output for Public Water Taps, Example

JSSP SANITATION SURVEY - PUBLIC WATER TAPS
ALL PUBLIC WATER TAPS
DATE: 04.22.1982

VAR.3/A AGE OF THE TAP

CATEGORY LABEL	CODE	ABSOLUTE FREQUENCY	RELATIVE FREQUENCY (PERCENT)	ADJUSTED FREQUENCY (PERCENT)	CUMULATED ADJ.FREQ. (PERCENT)
< 2 YEARS	0	48	37.1	37.1	37.1
2- 4 YEARS	1	59	46.8	46.8	83.9
5- 6 YEARS	2	16	12.9	12.9	96.8
7-10 YEARS	3	2	1.6	1.6	98.4
11-15 YEARS	4	2	1.6	1.6	100.0
TOTAL		127	100.0	100.0	

VAR.3/B OPERATION OF THE TAP

CATEGORY LABEL	CODE	ABSOLUTE FREQUENCY	RELATIVE FREQUENCY (PERCENT)	ADJUSTED FREQUENCY (PERCENT)	CUMULATED ADJ.FREQ. (PERCENT)
< 2 YEARS	0	12	9.5	9.5	9.5
2- 4 YEARS	1	13	10.2	10.2	19.7
5- 6 YEARS	2	2	1.6	1.6	21.3
NO OPERATION	6	100	78.7	78.7	100.0
TOTAL		127	100.0	100.0	

VAR.12 PRESSURE OF WATER

CATEGORY LABEL	CODE	ABSOLUTE FREQUENCY	RELATIVE FREQUENCY (PERCENT)	ADJUSTED FREQUENCY (PERCENT)	CUMULATED ADJ.FREQ. (PERCENT)
NO PRESSURE (NO WATER)	0	100	78.7	78.7	78.7
<0,5 BAR	1	13	10.2	10.2	88.9
0,5-1,0 BAR	2	12	9.5	9.5	98.4
1,1-1,5 BAR	3	1	0.8	0.8	99.2
1,6-2,0 BAR	4	1	0.8	0.8	100.0
TOTAL		127	100.0	100.0	

ANNEX 11

JSSP Sanitation Survey: Computer Output for MCK (Communal Sanitation Facilities), Example

JSSP SANITATION SURVEY - MCK
ALL MCK
DATE: 04.22.1982

VAR.4/A AGE OF MCK

CATEGORY LABEL	CODE	ABSOLUTE FREQUENCY	RELATIVE FREQUENCY (PERCENT)	ADJUSTED FREQUENCY (PERCENT)	CUMULATED ADJ.FREQ. (PERCENT)
2- 4 YEARS	1	6	25.0	25.0	25.0
5- 6 YEARS	2	2	8.3	8.3	33.3
7-10 YEARS	3	7	29.2	29.2	62.5
11-15 YEARS	4	5	20.8	20.8	83.3
>15 YEARS	5	4	16.7	16.7	100.0
TOTAL		24	100.0	100.0	

VAR.8/B CHARGES FOR TOILET - ADULT

CATEGORY LABEL	CODE	ABSOLUTE FREQUENCY	RELATIVE FREQUENCY (PERCENT)	ADJUSTED FREQUENCY (PERCENT)	CUMULATED ADJ.FREQ. (PERCENT)
< RP 5	0	3	12.5	14.3	14.3
RP 5 - RP 10	1	4	16.7	19.0	33.3
NO CHARGE	5	14	58.3	66.7	100.0
OUT OF RANGE		3	12.5	MISSING	
TOTAL		24	100.0	100.0	

VAR.12 STATE OF CLEANLINESS

CATEGORY LABEL	CODE	ABSOLUTE FREQUENCY	RELATIVE FREQUENCY (PERCENT)	ADJUSTED FREQUENCY (PERCENT)	CUMULATED ADJ.FREQ. (PERCENT)
KEPT ALWAYS CLEAN	0	12	50.0	55.0	55.0
SOMETIMES UNCLEAN	1	6	25.0	30.0	85.0
ALWAYS UNCLEAN	2	3	12.5	15.0	100.0
OUT OF RANGE		3	12.5	MISSING	
TOTAL		24	100.0	100.0	

VAR.23 NUMBER OF EMPLOYEES

CATEGORY LABEL	CODE	ABSOLUTE FREQUENCY	RELATIVE FREQUENCY (PERCENT)	ADJUSTED FREQUENCY (PERCENT)	CUMULATED ADJ.FREQ. (PERCENT)
1	0	8	33.3	38.0	38.0
2	1	5	20.9	24.0	62.0
NOBODY (MAINT.BY USERS)	2	8	33.3	38.0	100.0
OUT OF RANGE		3	12.5	MISSING	
TOTAL		24	100.0		

ANNEX 12

JSSP Sanitation Survey: Computer Output for Surface Drains, Example

JSSP SANITATION SURVEY - SURFACE DRAINS
ALL ZONES
DATE: 03.31.1982

VAR.P1 CONSTRUCTION MATERIAL

CATEGORY LABEL	CODE	ABSOLUTE FREQUENCY	RELATIVE FREQUENCY (PERCENT)	ADJUSTED FREQUENCY (PERCENT)	CUMULATED ADJ.FREQ. (PERCENT)
EARTH	0	7	4.6	4.6	4.6
STONE	1	22	14.8	14.8	19.4
CONCRETE	2	122	80.6	80.6	100.0
	TOTAL	151	100.0	100.0	

VAR.1 PHYSICAL CONDITIONS

CATEGORY LABEL	CODE	ABSOLUTE FREQUENCY	RELATIVE FREQUENCY (PERCENT)	ADJUSTED FREQUENCY (PERCENT)	CUMULATED ADJ.FREQ. (PERCENT)
GOOD	1	100	66.3	66.3	66.3
POOR	2	51	33.7	33.7	100.0
	TOTAL	151	100.0	100.0	

VAR.7 IS THE DRAIN CLEANED ?

CATEGORY LABEL	CODE	ABSOLUTE FREQUENCY	RELATIVE FREQUENCY (PERCENT)	ADJUSTED FREQUENCY (PERCENT)	CUMULATED ADJ.FREQ. (PERCENT)
YES	0	93	61.6	61.6	61.6
INADEQUATELY	1	58	38.4	38.4	100.0
	TOTAL	151	100.0	100.0	

VAR.11 HOW DOES THE DRAIN FUNCTION ?

CATEGORY LABEL	CODE	ABSOLUTE FREQUENCY	RELATIVE FREQUENCY (PERCENT)	ADJUSTED FREQUENCY (PERCENT)	CUMULATED ADJ.FREQ. (PERCENT)
VERY GOOD	0	99	65.6	65.6	65.6
CAPACITY TOO SMALL	1	19	12.6	12.6	78.2
SLOPE TOO FLAT	2	31	20.5	20.5	98.7
NO OUTLET POINT	3	2	1.3	1.3	100.0
	TOTAL	151	100.0	100.0	

ANNEX 13

JSSP SANITATION SURVEY: Repartition of Sub-areas for Extrapolation of Survey Results for House Sanitation Facilities

Sample Survey Zone		Results Extrapolable to Kampungs	Surface (ha)	Population (31.12.81)	Popul. Density (p/ha)
1	1 - Dukuh Setiabudi	1 - Dukuh Setiabudi	20	15,780[1]	789 [1]
		19 - Menteng Rawa Panjang	2	1,200	600
2	4 - Karet Belakang	2 - Karet Karya Utara	8	4,000	500
		3 - Karet Karya Selatan	4,25	3,417	804
		4 - Karet Belakang	30	12,487	416
		5 - Karet Belakang II	6	4,104	684
3	6 - Karet Pedurenan	6 - Karet Pedurenan	21	14,326	682
		8 - Karet Gang Mesjid	6	4,080	680
4	7 - Kuningan III	7 - Kuningan III	16	4,817	301
5	9 - Kuningan I	9 - Kuningan I	20	10,364	518
6	10 - Kuningan II	10 - Kuningan II	17	6,450	379
		11 - Karet Sawah/Depan	17	5,100	300
8	13 - Kawi Gembira	12 - Guntur	17,5	10,495	600
		13 - Kawi Gembira	12	5,400	450
9	14 - Menteng Wadas I	14 - Menteng Wadas I	17,5	16,406	937
		15 - Menteng Wadas II	15	7,977	531
10	16 - Menteng Atas	16 - Menteng Atas	30	16,506	550
11	17 - Menteng Rawa Panjang	17 - Menteng Rawa Panjang	24,25	19,114	788
12	18 - Kebon Obat	18 - Kebon Obat	10	6,250	625
	TOTAL SETIABUDI		293,5	168,273	573
10	16 - Menteng Atas	20 - Menteng Dalam Pal Batu	35	17,000	486
13	21 - Warung Pedok	21 - Warung Pedok	43	10,100	240
14	22 - Warung Pedok	22 - Warung Pedok II	52	15,200	292
15	24 - Manggarai Barat/T	24 - Manggarai Barat/T	29	21,938	756
16	25 - Bali Matraman	25 - Bali Matraman	64	32,957	514
17	27 - Bukit Duri Selatan	26 - Bukit Duri Puteran	17	10,742	632
		27 - Bukit Duri Selatan	26	7,000	269
		30 - Melayu Kecil	35	9,625	275
18	29 - Bukit Duri- Tanjakan	28E- Bukit Duri/Melayu Kecil	2	500	250
		29 - Bukit Duri Tanjakan	14	8,821	630
19	32 - Tebet Timur	28W- Bukit Duri/Melayu Kecil	2	500	250
		31 - Tebet Barat	100	25,937	259
		32 - Tebet Timur	115	26,123	227
		33 - Kebon Baru Kavling	20	5,000	250
		37 - Dalam Barat	9	1,810	201
20	36 - Kebon Baru	23 - Menteng Dalam Gang-Kober	4	800	200
		34 - Dalam Melayu Besar	18	10,270	570
		35 - Kebon Baru Utara	9	2,381	265
		36 - Kebon Baru	25	14,941	598
	TOTAL TEBET		619	221,645	358
	All Kampungs		912,5	389.918	427
	Non-Kampungs Areas	(without non-populated land 463 ha)	406,5	98,459	242
	TOTAL PROJECT AREA	(incl. non populated land)	1782	488,377	274

1) Incl. illegaly settled poeple (5000 - estimate)

ANNEX 14

JSSP SANITATION SURVEY: Repartition of Sub-areas for Extrapolation of Survey Results for Surface Drains

Kampungs : A - Low Income Areas, B - Medium Income Areas, C - High Income Areas

Group of Zones	Survey Results Extrapolable to Kampungs	Surface (ha)	Population (31.12.81)	Population Density (person/ha)
A	2 - Karet Karya Utara	8	4,000	500
	3 - Karet Karya Selatan	4,25	3,417	804
● 4 - Karet Belakang		30	12,487	416
	5 - Karet Belakang II	6	4,104	684
● 6 - Karet Pedurenan		21	14,326	682
	8 - Karet Gang Mesjid	6	4,080	680
● 9 - Kuningan I		20	10,364	518
	12 - Guntur	17,50	10,495	600
●13 - Kawi Gembira		12	5,400	450
●14 - Menteng Wadas I		17,50	16,406	937
	15 - Menteng Wadas II	15	7,977	531
●16 - Menteng Atas		30	16,506	550
●18 - Kebon Obat		10	6,250	625
	20 - Menteng Dalam Pal Batu	35	17,000	486
	23 - Menteng Dalam Gang Kober	4	800	200
●24 - Manggarai Barat/Timur		29	21,983	756
	28E - Bukit Duri/Melayu Kecil	2	500	250
●29 - Bukit Duri Tanjakan		14	8,821	630
	34 - Dalam Melayu Besar	18	10,270	570
	35 - Kebon Baru Utara	9	2,381	265
●36 - Kebon Baru		25	14.941	598
	Total A	333,25	192,463	577
B	●17 - Menteng Rawa Panjang	24,75	19,114	788
	●21 - Warung Pedok	43	10,100	240
	●22 - Warung Pedok II	52	15,200	292
	●25 - Bali Matraman	64	32,957	514
	●26 - Bukit Duri Puteran	17	10,742	632
	●27 - Bukit Duri Selatan	26	7,000	269
	30 - Melayu Kecil	35	9,625	275
	37 - Dalam Barat	9	1,810	201
	Total B	270,25	106,548	394
C	● 1 - Dukuh Setiabudi	20	15,780	789
	● 7 - Kuningan III	16	4,817	301
	●10 - Kuningan II	17	6,450	379
	11 - Karet Sawah/Depan	17	5,100	300
	19 - Menteng Rawa Panjang	2	1,200	600
	28W - Bukit Duri/Melayu Kecil	2	500	250
	31 - Tebet Barat	100	25,937	259
	●32 - Tebet Timur	115	26,123	227
	33 - Kebon Baru Kavling	20	5,000	250
	Total C	309	90,907	294
	GRAND TOTAL (A + B + C)	912,5	389,918	427
D	● NON-KAMPUNG AREAS (NON-POPULATED LAND) 1)	406,5 463,0	98,459 -	242 -
	TOTAL PROJECT AREA 2)	1782,0	488,377	274

Notes : 1) Non-populated and : open land, roads, rivers, lakes, etc.

2) Including small corner area between Sudirman and Subroto Highways and Krukut River (formaly excluded from the Project Area

● Sample Survey Zones

Kampong/Block	Number of Houses	Housing Pattern (%) 2				Distance of house to vehicular Road (%) 2					Rented Houses (%) 2	Rooms per House	Families per House	Occupants per House
		Permanent	Semi-Permanent	Temporary	Transient	<2.5m	25-50m	51-75m	76-100m	>100m				
1 - Rabah Setiabudi	2,254	79.0	11	3	1	99	1		-	-	24.5	5.2	1.1	7
2 - Karet Kuya Busa	608	25.0	58.8	13.8	2.5	53.7	28.7	5.0	-	17.5	30.0	3.9	1.1	6.5
3 - Karet Kuya Gubsan	519	25.0	58.8	13.8	2.5	53.7	23.7	5.0	-	17.5	30.0	3.9	1.1	6.5
4 - Karet Belakang	1,899	25.0	58.8	13.8	2.5	53.7	28.7	5.0	-	17.5	30.0	3.9	1.1	6.5
5 - Karet Bukbek II	624	25.0	58.8	13.8	2.5	53.7	23.7	5.0	-	17.5	30.0	3.9	1.1	6.5
6 - Karet Pedurenan	1,980	39.0	39.5	19.2	4.2	40.5	21.4	3.0	10.7	24.4	29.8	3.6	1.2	7.2
7 - Kuningan III	845	39.4	54.2	13.7	2.9	62.5	23.1	2.9	2.9	8.7	49.0	3.6	1.0	5.7
8 - Karet Gang Masdid	366	31.0	54.2	13.7	1.9	40.5	21.4	3.0	10.7	24.4	20.8	3.6	1.2	7.2
9 - Kuningan I	1594	44.9	43.0	10.3	1.9	54.4	17.8	7.5	1.5	21.5	18.7	3.9	1.0	6.5
10 - Kuningan II	1097	30.4	45.1	20.6	3.9	53.9	30.4	3.9	1.0	10.8	33.3	3.3	1.0	6.4
11 - Karet Samdh/Bapan	796	30.4	45.1	20.6	3.9	53.9	30.4	3.9	1.0	10.8	33.3	3.3	1.0	6.4
12 - Guntur	1478	31.3	56.3	10.0	2.6	30.5	35.0	5.0	7.5	15.0	23.7	3.7	1.1	7.1
13 - Kawi Gembira	760	31.3	56.3	10.0	2.6	30.5	35.0	5.0	7.5	15.0	23.7	3.7	1.1	7.1
14 - Menteng Wadas I	2604	35.3	52.9	5.9	5.9	67.6	22.1	2.0	2.9	4.4	35.3	3.1	1.1	6.3
15 - Menteng Wadas II	1266	35.3	52.9	5.9	5.9	67.6	22.1	2.0	2.9	4.4	35.3	3.1	1.1	6.3
16 - Menteng Atas	2500	50.5	33.3	14.3	1.9	57.1	22.9	8.6	9.5	1.9	27.6	3.8	1.1	6.6
17 - Menteng Rawa Panjang	3295	67.0	27.3	6.0	-	84.0	8.0	1.0	5.0	2.0	18.0	4.0	1.1	5.8
18 - Kebon Obat	946	47.7	40.2	12.1	-	28.0	29.0	9.3	12.1	21.5	18.7	3.5	1.1	6.6
19 - Menteng Rawa Panjang	171	7.9	1.1	3	1	99	1	-	-	-	24.5	5.2	1.1	7.0
Total Setiabudi	25721	44.0	41.9	122.1	2.0	56.4	21.2	4.5	5.4	12.6	26.4	3.8	1.1	6.7
20 - Menteng Dalam Pal Batu	2575	50.5	33.3	14.3	1.9	57.1	22.9	8.6	9.5	1.9	27.6	3.8	1.1	6.6
21 - Warung Pedok	1442	69.0	27.6	3.4	-	93.1	3.4	-	3.4	-	72.4	2.2	1.0	7.0
22 - Warung Pedok II	2375	67.7	27.7	4.6	-	52.3	24.6	9.2	7.7	6.2	21.5	3.7	1.0	6.4
23 - Menteng Da;am Gang Kober	114	63.6	22.7	12.1	1.5	28.8	33.3	7.6	22.7	7.6	21.2	4.1	1.0	7.0
24 - Manggarai Barat/Timur	3538	44.0	47.6	6.7	1.0	56.2	21.0	6.7	1.9	14.3	20.0	3.2	1.0	6.2
25 - Bali Matraman	4846	64.8	27.6	6.7	1.0	64.8	17.3	3.8	1.9	12.4	27.6	4.2	1.0	6.8
26 - Bukit Duri Puteran	1760	50.0	41.5	7.5	0.9	56.6	31.1	4.7	-	7.5	17.9	3.5	1.0	6.1
27 - Bukit Duri Selatan	1147	50.0	41.2	7.5	0.9	56.6	31.1	4.7	-	7.5	17.9	3.5	1.0	6.1
28W - Bukit Duri/Melayu Kecil	85	80.0	2.0	-	-	100.0	-	-	-	-	26.7	5.8	1.0	6.0
28E - Bukit Duri/Melayu Kecil	69	54.4	41.2	4.4	-	29.4	35.3	14.7	11.8	8.9	27.9	3.8	1.0	7.2
29 - Bukit Duri Tanjakan	1225	54.4	41.2	4.4	-	29.4	35.3	14.7	11.8	8.9	27.9	3.8	1.0	7.2
30 - Melayu Kecil	1577	50.0	41.5	7.5	0.9	56.6	31.1	4.7	-	7.5	17.9	3.5	1.0	6.1
31 - Tebet Barat	4322	80.0	20.0	-	-	100.0	-	-	-	-	26.7	5.8	1.0	6.0
32 - Tebet Timur	4353	80.0	20.0	-	-	100.0	-	-	-	-	26.7	5.8	1.0	6.0
33 - Kebon Baru Kavling	833	80.0	20.0	-	-	100.0	-	-	-	-	26.7	5.8	1.0	6.0
34 - Dalam Melayu Besar	1467	63.6	22.7	12.8	1.5	28.8	33.3	7.6	22.7	7.6	21.2	4.1	1.0	7.0
35 - Kebon Baru Utara	340	63.6	22.7	12.1	1.5	28.8	33.3	7.6	22.7	7.6	21.2	4.1	1.0	7.0
36 - Kebon Baru	2134	63.6	22.7	12.1	1.5	28.8	33.3	7.6	22.7	7.6	21.3	4.1	1.0	7.0
37 - Dalam Barat	301	80.0	20.0	-	-	100.0	-	-	-	-	26.7	5.8	1.0	6.0
Total Tebet	34501	56.7	35.1	7.5	0.7	52.6	25.0	7.1	5.9	9.5	22.3	3.9	1.0	6.4
All Kampungs	60222	48.3	39.6	10.6	1.6	55.1	22.4	5.4	5.5	11.6	25.0	3.8	1.1	6.6
Non Kampungs	13674	76.0	17.3	6.0	0.7	88.8	6.0	5.3	0.7	-	55.3	5.8	1.1	7.2
Total Project Area	73896	50.5	37.8	10.2	1.5	57.8	21.1	5.4	5.1	10.6	27.5	4.2	1.2	6.7

ANNEX 16
EXISTING WATER SUPPLY IN HOUSES (ESTIMATES)

Water Supply for Drinking and Cooking (%) of the total number of houses

KAMPUNG/AREA	Metered House Connection	Standpipe on plot	Water-tap/cistern Public	Water-tap/cistern Shared	Deepwell (>8m.) Public	Deepwell (>8m.) Shared	Deepwell (>8m.) Private	Handpump (KIP) Public	Handpump (<8 m), Shellow Well Private	Handpump (<8 m), Shellow Well Shared	Dugwell Private	Dugwell Shared	Venders	Other Sources
1 - Dakuh Setiabudi	1.1	1.1	–	–	–	–	64.9	–	28.7	–	4.3	–	–	–
2 - Karet karya Utara	3.7	2.5	–	–	–	6.3	25.0	2.5	13.7	1.2	26.2	16.2	–	2.5
3 - Karet Karya Selatan	3.7	2.5	–	–	–	6.3	25.0	2.5	13.7	1.2	26.2	16.2	–	2.5
4 - Karet Belakang	3.7	2.5	–	–	–	6.3	25.0	2.5	13.7	1.2	26.2	16.2	–	2.5
5 - Karet Belakang II	3.7	2.5	–	–	–	6.3	25.0	2.5	13.7	1.2	26.2	16.2	–	2.5
6 - Karet Pedurenan	–	–	1.8	–	1.8	8.9	16.1	5.4	41.1	6.0	6.0	14.3	–	0.6
7 - Kuningan III	–	–	1.0	–	1.0	4.8	17.3	3.8	11.5	2.9	35.6	23.1	–	–
8 - Karet Gang Mesjid	–	–	1.8	–	1.8	8.9	16.1	5.4	41.1	6.0	6.0	14.3	–	–
9 - Kuningan I	–	5.6	–	–	–	5.6	37.4	0.9	7.5	0.9	36.4	5.6	–	0.6
10 - Kuningan II	–	1.0	–	1.0	1.0	20.6	18.6	–	13.7	–	20.6	20.6	–	3.0
11 - Karet Sawah/Depan	–	1.0	–	1.0	1.0	20.6	18.6	–	13.7	–	20.6	20.6	–	3.0
12 - Gantur	–	6.3	4.0	2.5	–	1.2	16.0	1.2	51.2	3.7	3.7	12.5	3.7	3.7
13 - Kavl Gambira	–	6.5	24.0	2.5	–	1.2	6.0	1.2	51.2	3.7	3.7	12.5	3.7	3.7
14 - Menteng Wadas I	–	1.5	–	–	19.1	5.9	33.8	–	19.1	8.8	4.4	7.4	–	–
15 - Menteng Wadas II	–	1.5	–	–	19.1	5.9	33.8	–	19.1	8.8	4.4	7.4	–	–
16 - Menteng Atas	–	–	–	–	–	7.6	39.0	–	19.0	1.0	25.7	7.6	–	–
17 - Menteng Rawa Panjang	2.0	5.0	–	–	3.0	2.0	35.0	–	21.0	5.0	8.0	2.0	1.0	–
18 - Kebon Obat	–	–	–	–	–	1.9	46.7	–	29.9	5.6	5.6	10.3	–	16.0
19 - Menteng rawa Panjang	1.1	1.1	–	–	–	–	64.9	–	28.7	–	4.3	–	–	–
Total Setiabudi	0.5	1.9	1.0	0.3	1.9	6.2	29.6	1.5	24.0	3.2	16.1	11.1	0.4	2.3
20 - Menteng Dalam Pal Batu	–	–	–	–	–	–	39.0	–	19.0	1.0	25.7	7.6	–	–
21 - Wanang Pedok	12.1	20.0	3.0	–	–	7.6	47.2	–	3.4	–	10.8	3.4	–	–
22 - Wanang Pedok II	1.5	3.1	5.0	–	–	–	30.4	–	10.8	–	38.5	9.2	–	–
23 - Menteng Dalam Gang Kober	–	9.1	–	–	–	1.5	40.9	–	12.1	3.8	30.3	6.1	–	–
24 - Manggarai Barat/Timur	–	–	8.0	–	1.9	1.5	33.3	–	21.0	1.9	9.5	21.0	–	1.0
25 - Bali Matraman	7.5	5.7	0.9	–	2.9	8.6	30.1	–	22.9	–	22.9	1.9	–	–
26 - Bukit Duri Puteran	7.5	–	–	–	–	3.8	38.7	1.9	32.1	2.8	7.5	5.7	0.9	0.9
27 - Bukit Duri Selatan	7.5	–	–	–	–	0.9	38.7	1.9	32.0	2.8	7.5	5.7	0.9	0.9
28 - Bukit Duri/Melayu Kecil	35.0	20.0	–	–	–	0.9	45.0	–	–	–	–	–	–	–
29 - Bukit Duri/Melayu Kecil	13.2	4.4	–	–	2.9	1.5	22.1	–	20.6	2.9	8.8	1.5	1.5	20.6
29 - Bukit Duri Tanjakan	7.5	–	–	–	2.9	1.5	22.1	–	20.6	2.9	8.8	1.5	1.5	20.6
30 - Melayu Kecil	35.0	20.0	–	–	1.9	0.9	38.7	1.9	33.0	2.8	7.5	5.7	0.9	0.9
31 - Tebet Barat	35.0	20.0	–	–	–	–	45.0	–	–	–	–	–	–	–
32 - Tebet Timur	35.0	20.0	–	–	–	–	45.0	–	–	–	–	–	–	–
33 - Kebon Baru Kuning	–	9.1	–	–	–	–	45.0	–	–	–	–	–	–	–
34 - Bulan Melayu Besar	–	9.1	–	–	–	1.5	40.9	–	12.1	–	30.3	6.1	–	–
35 - Kebon Baru Utara	–	9.1	–	–	–	1.5	40.9	–	12.1	–	30.3	6.1	–	–
36 - Kebon Baru	–	9.1	–	–	–	1.5	40.9	–	12.1	–	30.3	6.1	–	–
37 - Bulan Barat	35.0	20.0	–	–	–	–	45.0	–	–	–	–	–	–	–
Total Tebet	3.4	3.0	0.2	–	1.2	3.0	37.4	0.4	20.3	2.0	18.2	7.7	0.4	2.9
All Kampungs	1.5	2.3	0.1	0.2	1.7	5.1	32.9	1.1	22.8	2.8	16.8	10.1	0.4	2.6
Non Kampungs Areas	52.7	1.3	–	–	–	–	9.3	–	14.0	1.3	9.3	11.3	0.7	–
Total Project Area	5.7	2.2	0.1	0.2	1.5	4.7	31.0	1.0	22.1	2.7	16.2	10.1	0.4	2.4

ANNEX 17
EXISTING TOILET SYSTEMS (Estimates)

Kampung/Area	None	Cistern Flush WC	Pour squat Plate	Pour squat Latrine	Ventilated Latrine	Unventilated Others	Unventilated Latrine	Neighbors MCK	Land	Open River	Drains	Others
1 – Dukuh Setiabudi	–	95.6	1.1	2.2	1.1	–	–	–	–	–	–	–
2 – Karet Karya Utara	–	86.6	12.4	–	1.0	–	–	–	–	–	–	–
3 – Karet Karya Selatan	–	86.6	12.4	–	1.0	–	–	–	–	–	–	–
4 – Karet Belakang	–	86.6	12.4	–	1.0	–	–	–	–	–	–	–
5 – Karet Belakang II	–	86.6	12.4	–	1.0	–	–	–	–	–	–	–
6 – Karet Pedurenan	2.5	73.3	22.7	–	1.5	–	4.6	–	73.92	1.5	20.0	–
7 – Kuningan III	7.0	68.5	17.0	4.0	3.5	–	100.0	–	–	–	–	–
8 – Karet Gang Mesjid	2.5	73.3	22.7	–	1.5	–	4.6	–	73.92	1.5	20.0	–
9 – Kuningan I	2.9	72.3	19.0	1.0	3.8	1.0	100.0	–	–	–	–	–
10 – Kuningan II	5.0	62.3	26.7	2.0	4.0	–	100.0	–	–	–	–	–
11 – Karet Sawah/Depan	5.0	62.3	26.7	2.0	4.0	–	100.2	–	–	–	–	–
12 – Guntur	34.7	43.0	13.9	–	4.2	4.2	26.3	2.6	68.42	2.6	–	–
13 – Karl Gembira	34.7	43.0	13.9	–	4.2	4.2	26.3	30.0	2.6	38.42	2.6	--
14 – Menteng Wadas I	29.4	63.0	2.9	1.5	3.2	–	5.0	90.0	5.0	–	–	--
15 – Menteng Wadas II	14.4	78.0	2.9	1.5	3.2	–	95.0	5.0	–	–	–	–
16 – Menteng Atas	1.9	83.6	11.8	–	2.9	–	100.0	40.0	–	–	–	–
17 – Menteng Rawa Panjang	6.1	77.5	3.1	3.1	2.0	8.2	10.0	–	90.22	20.0	20.0	10.0
18 – Kebon Obat	37.4	59.6	–	–	3.0	–	7.3	–	–	2.4	–	–
19 – Menteng Rawa Panjang	–	95.6	1.1	2.2	1.1	–	–	–	–	–	–	–
Total Setiabudi	16.9	64.9	12.9	1.2	3.0	1.1	25.9	10.8	0.9	53.3	2.4	6.6
20 – Menteng Dalam Pal Batu	1.9	83.6	11.6	–	2.9	–	100.0	–	–	–	–	–
21 – Warung Pedok	–	80.3	18.2	–	1.5	–	100.0	–	–	–	–	–
22 – Warung Pedok II	4.6	90.3	3.1	–	2.0	–	25.0	–	75.0	–	–	–
23 – Menteng Dalam Gang Kober	1.5	93.5	4.5	–	0.5	–	100.0	–	–	–	–	–
24 – Manggarai Barat/Timur	48.6	43.5	5.8	1.0	1.1	–	5.6	92.6	1.9	–	–	--
25 – Bali Matraman	1.9	87.5	10.6	–	–	–	50.0	–	50.0	–	–	–
26 – Bukit Duri Puteran	15.1	83.0	0.9	–	1.0	–	–	30.0	–	63.8	6.2	--
27 – Bukit Duri Selatan	15.1	83.0	0.9	–	1.0	–	–	30.0	–	63.8	6.2	--
28W – Bukit Duri/Melayu Kecil	–	100.0	–	–	–	–	–	–	–	–	–	–
28E – Bukit Duri/Melayu Kecil	4.8	88.6	3.2	1.6	1.8	–	12.5	–	87.5	–	–	–
29 – Bukit Duri Tanjakan	4.8	88.6	3.2	1.6	1.8	–	.12.5	–	87.5	–	–	–
30 – Melayu Kecil	15.1	83.0	0.9	–	1.0	–	–	30.0	–	63.8	6.2	–
31 – Tebet Barat	–	100.0	–	–	–	–	–	–	–	–	–	–
32 – Tebet Timur	–	100.0	–	–	–	–	–	–	–	–	–	–
33 – Kebon Baru Kavling	–	100.0	–	–	–	–	–	–	–	–	–	–
34 – Dalam Melayu Besar	1.5	93.5	4.5	–	0.5	–	100.0	–	–	–	–	–
35 – Kebon Baru Utara	1.5	93.5	4.5	–	0.5	–	100.0	–	–	–	–	–
36 – Kebon Baru	1.5	93.5	4.5	–	0.5	–	100.0	–	–	–	–	–
37 – Dalam Barat	–	100.0	–	–	–	–	–	–	–	–	–	–
Total Tebet	16.5	76.5	5.4	0.4	1.0	–	8.2	58.8	1.2	30.6	1.2	–
All Kampungs	16.8	69.2	10.3	0.9	2.1	0.7	20.9	24.6	1.0	46.8	2.0	4.7
Non Kampungs	0.7	99.3	–	–	–	–	50.0	–	50.0	–	–	–
Total Project Area	16.5	70.8	9.3	0.8	2.0	0.6	21.3	24.3	1.0	46.8	2.0	4.7

Toilet System in the house (% of the total number of hours)

If no latrine on the Not, the people ease (%)1

ANNEX 18

JSSP Sanitation Survey: Conditions of Existing Micro-drains in Setiabudi and Tebet Kecamatans

INCOME GROUP	AREA (ha)	LENGTH OF DRAINS m'	Construction Material			Physical Status		Age of Drains		
			EARTH m (%)	STONE m (%)	CONCRETE m' (%)	GOOD m' (%)	POOR m' (%)	0-6 YEARS m' (%)	6-10 YEARS m' (%)	MORE THAN 10 YRS m'(%)
A) LOW	333.25	191,308	3,443.5 (1.8)	20,278.7 (10.6)	167,585.8 (87.6)	143,481 (75.0)	47,827 (25.0)	148,027.4 (77.4)	26,209.2 (13.7)	16,643.8 (8.7)
B) MEDIUM	270.75	84,557.5	2,536.7 (3.0)	16,657.8 (19.7)	65,363 (77.3)	56,399.8 (66.7)	28,157.7 (33.3)	51,241.8 (60.6)	28,157.7 (33.3)	5,158 (6.1)
C) HIGH	309.0	123,936	4,957.6 (4.0)	17,351.0 (14.0)	101,627.4 (82.0)	89,233.8 (72.0)	34,702.2 (28.0)	94,191.3 (76.0)	9,914.9 (8.0)	19,828.8 (16.0)
NON KAMPUNG	406.5	124,326.5	14,297.5 (11.5)	23,870.7 (19.2)	86,158.3 (69.3)	57,438.8 (46.2)	66,887.7 (53.8)	38,292.6 (30.8)	57,438.8 (46.2)	28,595.1 (23.0)
TOTAL KAMPUNGS	913.0	399,801.5	9,995.0 (2.5)	53,573.4 (13.4)	336,233.1 (84.1)	289,856.1 (72.5)	109,945.4 (27.5)	292,654.7 (73.2)	69,565.5 (17.4)	37,581.3 (9.4)
TOTAL PROJECT AREA	1,782.5	524,128	24,292.5 (4.6)	77,444.1 (14.8)	422,391.4 (80.6)	347,294.9 (66.3)	176,833.1 (33.7)	330,947.3 (63.1)	127,004.3 (24.2)	66,176.3 (12.7)
NON POPULATED AREA	463	-	-	-	-	-	-	-	-	-

ANNEX 19
JSSP Sanitation Survey: Data Evaluation Related to Existing Public Water Taps

I t e m	Number	% 1)
1. Total number of Water taps	127	100
2. Water taps established by : Cipta Karya	17	13,4
Under KIP	104	81,9
Self-help of people	6	4,7
3. Planning/design criteria : population density (500 persons/tap)	119	93,5
water supply to small mosque	8	6,5
4. Age of the tap : 2 years	48	37,1
2 - 4 years	59	46,8
5 - 6 years	16	12,9
6 years	4	3,2
5. Operation of the tap : No operation	100	78,7
2 years	12	9,5
2 - 4 years	13	10,2
5 - 6 years	2	1,6
6. Average number of families taking water from the tap 2)	119	-
(Average number of persons using water from the tap) 3)	(714)	-
7. Average quantity of water taken from the tap 2) m3/day	9	-
8. Water from the tap used for2): drinking/cooking (m3/day)	3,6	40,0
other purposes (m3/day)	5,4	60,0
9. Hours in operation per day 2): 12 hours	5	18,0
17 hours	2	7,0
24 hours	20	75,0
10. Price of water sold at the tap 2): Rp 25/2 tins	8	30,0
No payment	19	70,0
11. The tap manager selected by 2) : Local Authorities	16	60,0
Warga (co-operative)	11	40,0
12. The tap is operated by 2): tap manager himself	16	60,0
person selected by the tap manager	11	40,0
13. The pressure of water : No water	100	79
0,5 bar	13	10,5
0,5 - 1,0 bar	12	9,3
1,0 - 1,5 bar	1	0,6
1,6 - 2,0 bar	1	0,6
14. Storage basin at the tap : None	123	96,3
Capacity : 1 m3	2	1,9
2 m3	2	1,9
15. Surveillance of the tap exercised by : RT/RW	110	87,1
No control	17	12,9
16. Physical status of the tap : very good	20	16,1
good	25	19,4
poor, because not in operation	82	64,5
17. Location of the tap : appropriate	115	90,3
not appropriate (within the private garden)	12	9,7

I t e m	Number	% 1)
18. The tap used primarily for : serving poor families	123	96,8
"private business"	4	3,2
19. Water does pass through the meter 2)	3	11,8
Water is by passing the meter 2)	24	88,2
20. Land furnished by : private owner	106	83,9
Municipality	21	16,1
21. Average number of families living in the service area :	136	-
thereof : poor families	126	92,5 4)
22. Average number of families in the service are using shallow wells	94	69,1 4)
These shallow wells were built : Under KIP	5	5,3
with self-help	61	64,9
by private contractors	28	29,8
Physical status of these wells : very good	3	3,5
good	43	45,6
poor	48	50,9

Notes : 1) Percentage of the total number of existing public water taps

2) Considered only taps in operation

3) Estimate based on the average number of persons per family in all Kampungs within the Project area (6 persons/family - see Annex 13)

4) Percentage of families living in the service area (136).

ANNEX 20

JSSP Sanitation Survey: Data Evaluation Related to Existing MCKs (Communal Sanitation Facilities)

I t e m	Number	% 1)
1. Total Number of MCKs	24	100,0
2. MCK established by : Public Works	1	4,2
Health Department	1	4,2
Under KIP	19	79,1
Self-help of the people	3	12,5
3. Planning/Design Criteria : One MCK per 1.000 persons	16	66,7
1001 - 2000 persons	8	33,3
4. Actual service area : 1000 persons	20	83,3
1001 - 3000 persons	4	16,7
5. Age of MCK : 2 - 4 years	6	25,0
5 - 6 years	2	8,3
7 - 10 years	7	29,2
11 - 15 years	5	20,8
15 years	4	16,7
6 MCK in operation : <u>No Operation</u>	3	12,5
2 - 4 years	6	25,0
5 - 6 years	2	8,3
7 - 10 years	7	29,2
11 - 15 years	5	20,8
15 years	1	4,2
7. Average number of families using MCK[2]: for water supply	65	
for toilet	53	
for bathing	52	
for washing	69	
8. Kind of Water Supply System : Deep wells with electric pump	7	29,0
Hand pump shallow well	14[3]	58,0
People bring water from house	3	17,0
9. Water Supplied to households [2] : m^3/day/MCK (average)	3,5	
10. Water used in MCK [2] - for toilets (m^3/day/MCK average)	0,8	
for bathing	2,5	
for washing	4,0	
11. Charges for MCK service [2] :		
a) for water supply : 40 Rp/2 tins (Rp 25)	2	9,5
No charge	19	90,5
b) for toilet : - adult : 5 Rp	3	14,3
5 - 10 Rp	4	19,0
No charge	14	66,7
- child : 5 Rp	2	9,5
No charge	15	90,5

I t e m	Number	% 1)
c) for bathing - adult : 10 Rp	2	9,5
10 - 15 Rp	1	4,8
No charge	18	85,7
- child : 5 Rp	2	9,5
No charge	19	90,5
d) for washing : 20 Rp	2	9,5
21 - 25 Rp	1	4,8
No charge	18	85,7
12. Physical status of MCK : good	14	58,0
poor, inadequate	10	42
13. State of cleanliness of MCK 2):-kept always clean	12	55,0
-sometime uncleaned	6	30,0
-always uncleaned	3	15,0
14. MCK uncleaned due to 4): MCK's staff insufficient	3	33,0
O & M not controlled	6	67,0
15. Status of the MCK Water Supply : very good	2	8,3
good	9	37,5
poor inadequate	13	54,2
16. Poor status of repair due to 5): water supply very old	5	39,5
maintenance neglected	1	12,0
Lack of financial-sources	7	49,5
17. Daily time of operation 2): 10 hours	2	10,0
20 hours	4	19,0
24 hours	15	71,0
18. Manager of the MCK selected 2) : by PAM	1	5,0
by RW/RT	18	85.0
by co-operation of people	2	10,0
19. Operator2): The Manager himself	7	33,0
Person selected by the Manager	13	62,0
By co-operation of people	1	5,0
20. Toilet system in the MCK :		
- WC connected to septic tank with leaching system	9	37,5
- WC connected to septic tank only	5	20,8
- Water seal with pour flush	4	16,7
- Squatting plate with pour flush	6	25,0
21. MCK with 4 toilet units	6	25,0
6 - " -	5	20,8
8 - " -	8	33,3
10 - " -	3	12,5
12 - " -	1	4,2
16 - " -	1	4,2
22. MCK with 2 laundry units (chambers)	14	58,4
4 - " -	4	16,7
6 - " -	1	4,2
8 - " -	5	20,8

I t e m	Number	% 1)
23. MCK with 2 bathing units (chambers)	19	59,1
6 - " -	1	4,2
8 - " -	3	12,5
12 - " -	1	4,2
24. Number of employees 2) : 1	8	38,0
2	5	24,0
Nobody	8	38,0
25. Average number of poor families in the MCK service area	118	
Thereof : families using MCK	98	83 6)
Other families 6) use : friends facilities	7	6
river	13	11
26. Families using MCK have to walk 7): 25 m	25	25
26 - 50 m	44	45
51 - 100 m	29	30
27. Maximum destance people willing to walk 6) : 50 m		47
100 m		43
150 m		5
200 m		5
28. Land for MCK furnished by : Private owner	11	45,8
Municipality	9	37,5
Co-operative land	4	16,7
This has influenced : - size and location of MCK	5	20,8
- selection of the MCK Manager	5	20,8
29. Did the MCK meet its intended objective ?		
- Yes	10	41,7
- No, because of small size	6	25,0
lack of privacy	1	4,2
O & M neglected	1	4,2
lack of water	4	16,7
combined above reasons	2	8,3

Notes : 1) Percentage of the total number of MCK (24)

2) Percentage of the total number of MCKs in operation (21)

3) Three MCKs have a water tap, but out of operation (no water)

4) Percentage of the total number of MCKs uncleaned (9)

5) Percentage of the total number of MCKs with poor water supply (13)

6) Percentage of poor families in the MCK service area (118)

7) Percentage of families using MCK (98)

ANNEX 21

IMPROVEMENT PROPOSALS : Public Water Taps

Kampung	Area (ha)	Total Population (31.12.1981)	Popul. using water for drinking & cooking from shallow wells & dug wells	Water Source — PAM Mains	Water Source — KIP Deep Wells	Existing — In Operation	Existing — Out of Operation (no water)	Existing — Total	Rehabilitation of Taps	New Tap to be built 7)	Distribution Network be built (72 m/ha)	Unit Cost — Rehab of Tap	Unit Cost — New taps 7)	Unit Cost — Pipes per m	Total Cost — Rehabilitation	Total Cost — New Taps	Total Cost — Pipes	Grand Total
1 – Dukuh Setiabudi	20	15,780 5)	5,200 5)	x	-	1	-	1	-	3	430 10)	15,0 9)	131,0	8,0	-	393	3440	3833
2 – Karet Karya Utara	8	4,000	2,500	-	-	-	-	-	-	2	-				-	262	-	262
3 – Karet Karya Selatan	4,25	3,417	2,139	x	-	-	-	-	-	8	-				-	1084	-	1084
4 – Karet Belakang	30	12,487	7,780	x	-	-	5	5	5	-	-				75	-	-	75
5 – Karet Belakang II	6	4,104	2,560	x	x 2)	-	5	5	5	-	-				75	-	-	75
6 – Karet Pedurenan	21	14,326	11,300	x	-	-	3	3	3	10	-				45	1310	-	1355
7 – Kuningan III	16	4,817	3,700	-	x 2)	-	5	5	5	-	-				75	-	-	75
8 – Karet Gang Mesjid	6	4,080	3,200	-	-	-	3	3	3	1	-				45	131	-	176
9 – Kuningan I	20	10,364	5,300	x	-	-	4	4	4	4	-				60	524	-	584
10 – Kuningan II	17	6,450	3,735	x	-	-	3	3	3	2	-				45	262	-	307
11 – Karet Sawah/Depan	17	5,100	2,950	x	-	-	-	-	-	-	-				-	-	-	-
12 – Guntur	17,5	10,495	8,360	x	x 3)	1	-	17 6)	-	8 8)	300 10)				-	1048	2400	3448
13 – Kawi Gembira	12	5,400	3,220	-	x 3)	13	4	25	4	-	-				60	-	-	60
14 – Menteng Wadas I	17,5	16,406	6,510	x	x 4)	-	25	25	25	-	-				375	-	-	375
15 – Menteng Wadas II	15	7,977	3,170	x	-	-	-	-	-	3	-				-	393	-	393
16 – Menteng Atas	30	16,506	8,800	x	x 4)	-	6	6	6	9	2160				90	1179	17280	18459
17 – Menteng Rawa Panjang	24,25	19,114	10,130	x	x 4)	-	5	5	5	7	430 10)				75	917	3440	4447
18 – Kebon Obat	10	6,250	3,215	x	-	-	6	6	6	-	-				90	-	-	75
19 – Menteng Rawa Panjang	2	1,200	400	x	-	-	5	5	5	-	-				75	-	-	-
Total Setiabudi	293,5	168,273	94,160	13	7	15	63	78	63	57	3320				945	7503	26560	34008
20 – Menteng Dalam Pal Batu	35	17,000	9,060	x	-	-	-	-	-	9	2520				-	1179	20160	21339
21 – Warung Pedok	43	10,100	1,780	x	-	1	1	1	-	1	-				-	131	-	131
22 – Warung Pedok II	52	15,200	8,900	x	-	2	4	6	4	6	-				60	786	-	846
23 – Menteng Dalam Gang Kober	4	800	390	x	-	-	-	-	-	-	-				-	-	-	-
24 – Manggarai Barat/Timur	29	21,938	12,350	x	-	8	10	10	10	2	-				150	262	-	412
25 – Bali Matraman	64	32,957	16,150	x	-	1	11	19	11	-	-				165	-	-	165
26 – Bukit Duri Puteran	17	10,742	5,560	x	-	-	4	5	4	3	860 10)				60	393	6880	7333
27 – Bukit Duri Selatan	26	7,000	3,690	x	-	-	4	-	4	4	130 10)				60	524	10720	11244
28 – Bukit Duri/Melayu Kecil	2	500	-	-	-	-	-	-	-	1	140				-	131	1120	1251
28 – Bukit Duri/Melayu Kecil	2	500	360	-	-	-	6	6	6	-	-				90	-	-	90
29 – Bukit Duri Tanjakan	14	8,821	4,930	x	-	-	-	-	-	1	-				-	131	-	-
30 – Melayu Kecil	35	9,625	5,070	x	-	1	-	1	-	5	2520				-	655	20160	20815
31 – Tebet Barat	100	25,937	-	x	-	-	-	-	-	-	-				-	-	-	-
32 – Tebet Timur	115	26,123	-	x	-	-	-	-	-	-	-				-	-	-	-
33 – Kebon Baru Kavling	20	5,000	4,980	x	-	-	-	-	-	5	-				-	655	-	655
34 – Dalam Melayu Besar	18	10,270	1,150	x	-	-	1	1	1	1	-				15	131	-	-
35 – Kebon Baru Utara	9	2,381	-	x	-	-	2	2	2	2	300				30	986	2400	2546
36 – Kebon Baru	25	14,941	7,250	x	-	-	4	4	4	6	1000				60	-	8000	9016
37 – Dalam Barat	9	1,811	-	-	-	-	-	-	-	-	-				60	-	-	60
Total Tebet	619	221,645	81,620	17	-	12	37	49	37	43	8680				630	5833	69440	75903
All Kampungs	912,5	389,918	175,780	29	7	27	100	127	100	100	12000	-	-	-	1575	13336	96000	110911 1)

Notes :
1) Cost level : June 1982 (Basic Cost)
2) Deep Well serving both Kampungs (No. 6 and 8); a new one with electric pump but not yet connected to the electricity
3) Deep Well serving both Kampungs (No. 12 and 13); deep well with electric pump in operation (Deep Well of PAM)
4) Deep Well provided with electric pump (connected to the electricity, but not yet in operation)
5) Including illegal settlers in the area (5000-estimate)
6) Non-KIP water taps (PAM)
7) Planning Criteria : 1000 persons/tap (type A – 2 faucets)
8) Settlements of Kampung 11 – Karet Sawah/Depan behind the river Krukut (using shallow wells & dug well) expected to be moved for high building
9) Fitting crans and connections to pipes
10) Extension of distribution system.

ANNEX 22

IMPROVEMENT PROPOSALS: MCKs

Kampung	Area (ha)	Total Population (31.12.1981)	Population without toilet system in house	Number of Existing MCKs — In Operation	Out of Operation	Total	MCK be rehabilitated covered & extended/seats	New MCK be established /Type 1)	Unit cost (Rp 1000) 2) Rehabilit./Roofing, Extention	New MCK #)	Total Cost (Rp 1000) Rehabilitation Extension	New MCKs	Grand Total 2)
1 - Dukuh Setiabudi	20	15,780	-				-						
2 - Karet Karya Utara	8	4,000	-										
3 - Karet Karya Selatan	4,25	3,417	-										
4 - Karet Belakang	30	12,487	-										
5 - Karet Belakang II	6	4,104	-										
6 - Karet Pedurenan	21	14,326	360					1/B		3,000		3,000	3,000
7 - Kuningan III	16	4,817	340					1/B		3,000		3,000	3,000
8 - Karet Gang Mesjid	6	4,080	105					1/A		1,800		1,800	1,800
9 - Kuningan I	20	10,364	300					1/B		3,000		3,000	3,000
10 - Kuningan II	17	6,450	325					1/B		3,000		3,000	3,000
11 - Karet Sawah/Depan	17	5,100	255					- 3)					
12 - Guntur	17,5	10,495	3640				1/32 seats 4)	3/C	R-Rf-E 8)	3,600	17,170	10,800	27,970
13 - Kawi Gembira	12	5,400	1875	5	1	6	3/-	5/B	750,0	3,000	2,250	15,000	17,250
14 - Menteng Madas I	17,5	16,406	4825	3	-	3		2/B		3,000		6,000	6,000
15 - Menteng Madas II	15	7,977	2350					1/B		3,000		3,000	3,000
16 - Menteng Atas	30	16,506	315		1 5)	1		-					
17 - Menteng Rawa Panjang	24,25	19,114	1170	3	-	3	3/-	3/B	750,0	3,000	2,250	9,000	11,250
18 - Kebon Obat	10	6,250	2340	1	-	1	1/-	-	750,0		750		750
19 - Menteng Rawa Panjang	2	1,200	-										
Total Setiabudi	**293,5**	**168,273**	**18200**	**12**	**2**	**14**	**8/32 seats**	**19**			**22,420**	**57,600**	**80,020**
20 - Menteng Dalam Pal Batu	35	17,000	325					1/B		3,000		3,000	3,000
21 - Marung Pedok	43	10,100	-										
22 - Marung Pedok II	52	15,200	700					2/B		3,000		6,000	6,000
23 - Menteng Dalam Gang Kober	4	800	10										
24 - Manggarai Barat/Timur	29	21,938	10660	6	1	7	1/12 6)	1/B 7)	R - Rf 8)	3,000	5,320	3,000	8,320
25 - Bali Matraman	64	32,957	630					2/B		3,000		6,000	6,000
26 - Bukit Duri Puteran	17	10,742	1620	1	-	1	1/-	1/B	750,0	3,000	750	3,000	3,750
27 - Bukit Duri Selatan	26	7,000	1060	1	-	1	1/-	1/B	750,0	3,000	750	3,000	3,750
28 - Bukit Duri/Melayu Kecil	2	500	20										
28E - Bukit Duri/Melayu Kecil	2	500	-										
29 - Bukit Duri Tanjakan	14	8,821	425	1	-	1	1/-	1/B	750,0	3,000	750	3,000	3,750
30 - Melayu Kecil	35	9,625	1450					1/B		3,000		3,000	3,000
31 - Tebet Barat	100	25,937	-										
32 - Tebet Timur	115	26,723	-										
33 - Kebon Baru Kaviling	20	5,000	-										
34 - Dalam Melayu Besar	18	10,270	150					1/A		1,800		1,800	1,800
35 - Kebon Baru Utara	9	2,381	30					1/A		1,800		1,800	1,800
36 - Kebon Baru	25	14,941	220										
37 - Dalam Barat	9	1,811	-										
Total Tebet	**619**	**221,645**	**17300**	**9**	**1**	**10**	**4/12-**	**12**			**7,570**	**33,600**	**41,170**
All Kampungs	**912,5**	**389,918**	**35500**	**21**	**3**	**24**	**12/20 - 32**	**31 (3/A, 25/B, 3/C)**			**29,990**	**91,200**	**121,190**

Notes :
1) Types : A - 4 toilet units (seats), B - 8 toilet units, C - 12 toilet units
2) Costs level : June 1982 (without land)
3) Settlements outside the Project Area (part expected to move within 5 years for high rised building)
4) Rehabilitation of 1 MCK (8 seats) and extension of 5 MCK (32 seats) and roofing existing MCK
5) Destroyed and on the site built an office
6) Rehabilitation of 1 MCK (12 seats) and roofing existing MCK
7) Be established in K - 24 east (to replace the destroyed one)
8) Rehabilitation (R) = Rp. 70.000/MCK, Roofing (Rp) = Rp 750.000/MCK, Extension (E: = see note 9)
9) Type A = Rp 1,800.000, Type B = Rp 3,000.000, Type C = Rp 3,600.000

ANNEX 23
IMPROVEMENT PROPOSALS : Leaching Pits (L.P.)

Kampung	Total Number of Houses	Houses with L.P.		New Leaching Pits Required 1)	Cost Estimate (June 1982) Rp 1000 2)
		Total Number	Thereof built under KIP		
1 - Dukuh Setiabudi	2,254	622	-	25	
2 - Karet Karya Utara	608	410	-	7	
3 - Karet Karya Selatan	519	350	-	6	
4 - Karet Belakang	1,899	1,282	-	23	
5 - Karet Belakang II	624	421	-	7	
6 - Karet Pedurenan	1,989	1,644	1,306	189	
7 - Kuningan III	845	682	-	49	
8 - Karet Gang Mesjid	566	468	372	54	
9 - Kuningan I	1,594	1,275	-	30	
10 - Kuningan II	1,007	770	-	20	
11 - Karet Sawah/Depan	796	609	-	16	
12 - Guntur	1,478	406	-	331	
13 - Kawi Gembira	760	209	-	170	
14 - Menteng Wadas I	2,604	1,263	-	192	
15 - Menteng Wadas II	1,266	614	-	93	
16 - Menteng Atas	2,500	1,690	19	25	
17 - Menteng Rawa Panjang	3,295	922	-	(x) 923	
18 - Kebon Obat	946	443	-	18	
19 - Menteng Rawa Panjang	171	47	-	-	
Total Setiabudi	25,721	14,127	1,697	2,178	544,500
20 - Menteng Dalam Pal Batu	2,575	1,740	-	26	
21 - Warung Pedok	1,442	196	-	-	
22 - Warung Pedok II	2,375	1,059	-	(x) 218	
23 - Menteng Dalam Gang Kober	114	79	-	-	
24 - Manggarai Barat/Timur	3,538	1,079	1	134	
25 - Bali Matraman	4,846	2,767	25	-	
26 - Bukit Duri Puteran	1,760	814	-	132	
27 - Bukit Duri Selatan	1,147	531	103	86	
28W - Bukit Duri/Melayu Kecil	83	6	-	-	
28E - Bukit Duri/Melayu Kecil	69	31	-	4	
29 - Bukit Duri Tanjakan	1,225	558	-	72	
30 - Melayu Kecil	1,577	730	-	118	
31 - Tebet Barat	4,322	289	-	-	
32 - Tebet Timur	4,353	291	-	-	
33 - Kebon Baru Kavling	833	56	-	-	
34 - Dalam Melayu Besar	1,467	1,022	-	-	
35 - Kebon Baru Utara	340	237	-	-	
36 - Kebon Baru	2,134	1,487	-	-	
37 - Dalam Barat	301	20	-	-	
Total Tebet	34,501	12,992	129	790	197,500
All Kampungs	60,222	27,119	1,826	2,968	742,000 3)
Thereof : Pilot Project				40	10,000

Notes : 1) For toilet system of which water used is going to drains, ditches, rivers or open land
2) Basic Cost estimate, based on an average unit cost of Rp 250.000,- per leaching pit (lined with concrete)
3) Government subsidies considered (5% of the total cost)
x) Areas selected for Pilot Project (20 leaching pits each)

ANNEX 24

IMPROVEMENT PROPOSALS : Surface Drains

Area	Surface (ha)	Length of Drains (m)	Repairing of existing drains (m) 2)	Construction of drains (m) 3)	Improvement Proposals Cost estimate (Rp 1000) 4)		
					Repairing	Construction	Total
A) KAMPUNG AREAS :							
- Setiabudi	293,5	149,131,5	41,011	3,728	32,809	59,648	92,457
- Tebet	619,0	250,670,0	68,934	6,267	55,147	100,272	155,419
Total	912,5	399,801,5	109,945	9,995	87,956	159,920	247,876
B) NON-KAMPUNG AREAS :	406,5	124,326,5	66,888	14,297	53,510	228,752	282,262
Non-Populated Area 1)	463,0	-	-	-	-	-	-
C) TOTAL PROJECT AREA	1,782,0	524,128	176,833	24,292	141,466	388,672	530,138

1) Non-Populated Area : Open land, roads, railway, rivers, lakes, etc.

2) Drains of which physical status is poor - see Annex 33, Page 3

3) Earth drains - see Annex 33, Page 3

4) Basic Cost Estimate (level : June 1982) based on the following average unit cost :

   - Rp 16.000/m for construction of new drains
   - Rp   800/m for repairing of existing drains (5% of the capital cost)

ANNEX 25

JSSP: CAPITAL COST ESTIMATES OF PROPOSED SANITATION IMPROVEMENTS - SUMMARY

Item	June 1982 Cost Estimate (in Million Rupiah)				
	Basic Cost	Physical 1) Contingencies	Price 2) Contingencies	Engineering 3)	Total
1. Public Water Taps:					
Rehabilitation (100 taps)	1.6	0.5	0.5	0.4	3.0
New taps (100 taps)	13.4	3.5	3.5	3.0	23.4
Extension of pipes network (1,200 m)	96.0	24.0	24.0	21.6	165.6
Deep well stations (8)	288.0	72.0	72.0	65.0	497.0
Land acquisition (240 m2)[4]	12.0	-	2.0	-	14.0
TOTAL (a)	411.0	100.0	102.0	90.0	703.0
2. MCKS:					
Rehabilitation, including roofing of existing MCKs	30.0	7.5	7.5	6.8	51.8
New MCKs (31 with a total of 248 toilets)	91.2	22.8	22.8	20.5	157.3
Land acquisition (3,100 m2)[4]	155.0	-	31.0	-	186.0
TOTAL (b)	276.2	30.2	61.3	27.3	395.1
3. Leaching Pits and Desludging					
Pilot project (40 pits)	10.0	2.5	2.5	2.3	17.3
Government subsidies (for 2,928 pits)[5]	36.6	9.2	9.2	8.3	63.3
Desludging equipment	162.0	40.5	40.5	36.4	279.4
Transfer and thickening station (TTS)	62.4	15.6	15.6	14.0	107.6
Land acquisition for TTS (1,100 m2)	55.0	-	11.0	-	66.0
TOTAL (c)	326.0	67.8	78.8	61.0	533.6
4. Surface Drains:					
Repairing of existing drains (176.8 km)	141.5	35.4	35.4	31.8	244.1
Construction of drains (24.3 km)	388.7	97.2	97.2	87.5	670.6
TOTAL (d)	530.2	132.6	132.6	119.3	914.7
GRAND TOTAL (a + b + c + d)	1,543.4	330.7	374.7	297.6	2,546.4

Notes: 1) 25 % of basic costs
2) 20 % of basic costs and physical contingencies
3) 15 % of basic costs and contingencies
4) Unit cost: Rp 50.000/m2
5) 5 % of the total cost of leaching pits

ANNEX 26

Breakdown of Cost Estimate of Proposed Improvements (Million Rupiah)

Item	Local Cost	Foreign Cost	Total
1. Public Water Taps:			
Rehabilitation of existing taps	1.6		1.6
New taps	13.4		13.4
Extension of pipes network	38.4	57.6	96.0
Deep well stations	224.0	64.0	288.0
Land acquisition	12.0	−	12.0
TOTAL (a)	289.4	121.6	411.0
2. MCKs:			
Rehabilitation of existing MCKs	30.0	−	30.0
New MCKs	91.2	−	91.2
Land acquisition	155.0	−	155.0
TOTAL (b)	276.2	−	276.2
3. Leaching Pits and Desludging:			
Pilot project	10.0	−	10.0
Government subsidies	36.6	−	36.6
Desludging equipment	−	162.0	162.0
Transfer and thickening station	53.3	9.1	62.4
Land acquisition for TTS	55.0	−	55.0
TOTAL (c)	154.9	171.1	326.0
4. Surface Drains:			
Repairing of existing drains	134.3	7.2	141.5
Construction of drains	369.1	19.6	388.7
TOTAL (d)	503.4	26.8	530.2
Total Basic Costs (a + b + c + d)	1,223.9	319.5	1,543.4
Physical Contingencies	250.7	80.0	330.7
Price Contingencies	294.8	79.9	374.7
Engineering	59.6	238.0	297.6
GRAND TOTAL	1,829.0	717.4	2,546.4

ANNEX 27

JSSP – SANITATION PROJECT : IMPLEMENTATION SCHEDULE 1)

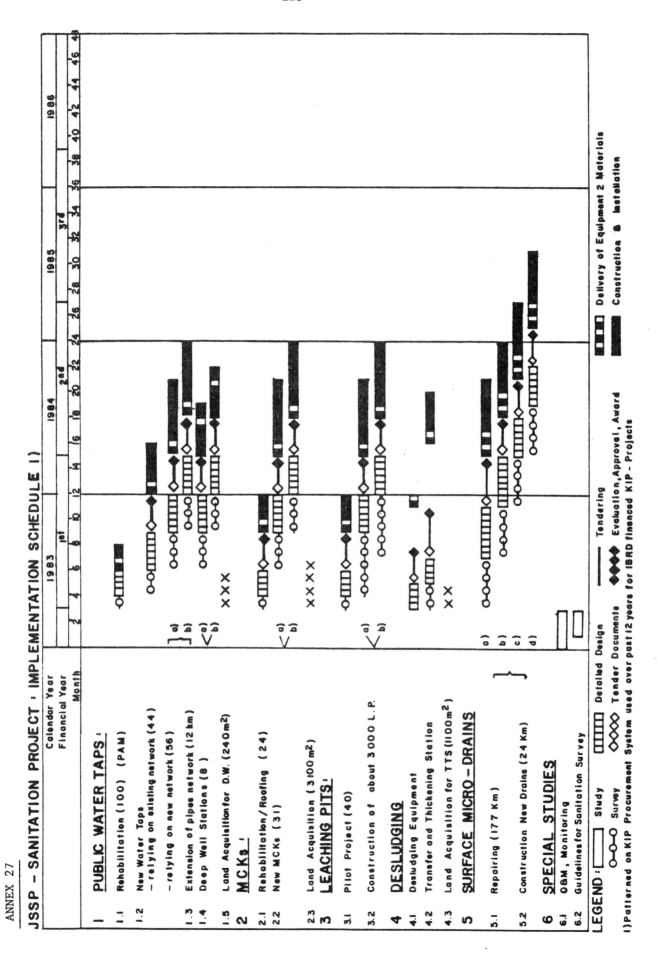

ANNEX 28

TERMS OF REFERENCE FOR IMPLEMENTATION OF SANITATION PROJECT

---

1   DETAILED ENGINEERING

As a part of the detailed engineering was already accomplished within the feasibility study (particularly with typical designs and standards of proposed facilities), this work shall be focused on the location of facilities and appropriate adaptation of typical designs to the sites selected.

1.1   WATER SUPPLY

The detailed engineering for water supply will comprise the following work:

1.1.1   Rehabilitation of Existing Public Water Taps

a)   to identify the deficiencies of all public water taps out of operation (100), such as missing taps, meters, disconnections from the distribution network, etc.

b)   to describe the work to be done for each tap, with specification of material and labor requirements, as well as estimates of the time needed and cost of repairs

c)   to establish a plan (time schedule) for the supply of materials, fitting of equipment and implementation of repairs.

1.1.2   New Water Taps, Deep Wells and Extension of Distribution Network

a)   to carry out in all kampungs concerned a detailed field survey required for the detailed design of the distribution network to be extended (12,000 m) as well as for the location of new public water taps (100)

b)   to carry out in all kampungs concerned a detailed soil and ground water survey required for the detailed design and location of new deep well stations (8); in this connection to arrange for physical, chemical and bacteriological analysis of water samples

c) to prepare, in close cooperation with PAM DKI[1], a detailed engineering design for the distribution network to be extended, for new deep wells, as well as for new public water taps, whereby existing KIP[2] standards, specifications and planning criteria could be applied

d) to propose procedures for procurement and prepare appropriate tender documents for bidding whereby only local procurement will be considered; tender documents should comprise:

- tender notices and invitations
- instructions for tendering
- drawings
- general conditions
- technical specifications
- proposal forms
- construction schedule
- bill of quantities
- equipment lists

e) to prepare a list of materials which may be offered locally and draw up appropriate specifications

f) to prepare proposals for contract packages

g) to assist the Client in the evaluation of bids submitted

h) to prepare and submit to the Client the detailed engineering report which should also include a time schedule for the implementation of work.

1.1.3 Distribution System Analysis

a) to carry out a detailed engineering analysis of the existing distribution system and make proposals for improvements in the system necessary to reduce the extend of unaccounted water and losses and to increase the efficiency of the system

b) to carry out leakage detection surveys after commissioning new facilities.

---

1) PAM DKI = Perusahaan Air Minum DKI (Jakarta Water Supply Company)

2) KIP = Kampung Improvement Program

1.2   MCKs[1]

1.2.1   Rehabilitation of Existing MCKs

a) to identify the deficiencies of all existing MCKs (24), particularly those out of operation and the work to be done for each MCK with specification of material and labor requirements, as well as estimates of time needed and cost of repairs

b) to carry out a detailed engineering investigation and design for roofing all existing MCKs, whereby a simple low-cost structure (allowing ventilation and light) is to be considered

c) to establish a plan (time schedule) for the supply and testing of materials and implementation of repairs and roofing.

1.2.2   New MCKs

a) to carry out in all kampungs concerned a detailed field survey required for the detailed design and location of new MCKs (31); the field survey will be focused on land availability, water supply, accessibility of site for desludging septic tank, characteristics of soil and environmental aspects

b) to prepare detailed engineering design for new MCKs, whereby existing KIP standards of roofed MCKs, specifications and planning criteria could be applied .

c) to propose procedures for procurement and prepare appropriate tender documents for bidding (with similar contents as for water supply - see Paragraph 1.1.2) whereby only local procurement will be considered

d) to prepare lists of materials, appropriate specifications and contract packages

e) to assist the Client in the evaluation of bids submitted and prepare the detailed engineering report.

---

1) MCK = mandi-cuci-kakus (Communal Sanitation Facility, including Toilets, Bathing and Washing Units)

1.3 LEACHING PITS AND DESLUDGING

The detailed engineering for individual building of excreta
disposal systems and related desludging activities will comprise
the following scope of work:

1.3.1 Pilot Project for Leaching Pits

a) to carry out, in close cooperation with local authorities
(Kecamatans, Kelurahans, RW, RT)[1], a detailed field
investigation and appropriate consultations with families
concerned in Kampung 17 - Menteng Rawa Panjang and Kampung 22
- Warung Pedok II, in order to provide all data necessary for
the detailed design of new (40) leaching pits (20 per
selected zone) and to obtain the consent of family heads for
construction

b) to prepare a detailed engineering design for the above 40
leaching pits (lined with concrete), whereby larger capacity
pits must be considered

c) to propose procedures for procurement and prepare appropriate
tender documents for bidding, whereby only local procurement
will be considered

d) to prepare lists of materials, appropriate specifications and
contract packages

e) to assist the Client in the evaluation of bids submitted and
prepare the detailed engineering report

f) after 1 year's operation of the 40 leaching pits built within
the pilot project, to assist the Client:

   (i)   in evaluation of experience gained from their
          construction and operation

   (ii)   in preparation of new regulations related to house
          sanitation facilities

   (iii) in preparation of a campaign to promote the building
          of the remaining leaching pits (about 3,000) by
          householders themselves, as well as

   (iv)   in preparation of a detailed financing scheme for the
          above construction of leaching pits by householders,
          whereby government incentives such as soft government
          credits, differentiated subsidies, technical assis-
          tance, etc., could also be considered. The extent of
          subsidies and other financial incentives shall be
          determined, in each particular case, in consultation

1) RW (Rukun Warga) and RT (Rukun Tetangga) = basic administrative units - local representatives selected by the population

with a committee consisting of representatives of
local authorities and population

In this connection, in order to achieve total excreta
management, ordinance(s) needed to suit the new situation
should be based on the following principles:

(i)   all buildings close enough to sewers must be forced to
        connect and to pay a connection fee; for poor people's
        houses, the government could permit payment by
        installments over a long period

(ii)   for houses in kampungs not in (i) above, it must be
        required to have adequate leaching pits and to
        desludge them as frequent as necessary. The government
        is expected to provide standards for acceptable pit
        design and requirements on desludging, and to help
        poor families with payment, whereby two approaches
        could be considered: the government builds and col-
        lects money or the families are forced to build with
        soft government credits

(iii)   for persons not having toilets in houses nor access to
        them, it must be required to use MCKs provided by the
        government.

1.3.2   Sludge Transfer and Thickening Station

a) to prepare specifications for soil investigation

b) to arrange for soil investigation by local subcontract

c) to check results of campaign and formulate conclusions

d) to carry out detailed topographical mapping of selected
    sites

e) to elaborate detailed design including:

    - architectural elevations
    - site plan, access roads, drainage
    - structural analysis
    - formwork and reinforcement plans
    - working drawings for superstructure
    - tender design for electrical and mechanical equipment
    - office, workshop and staff housing

f) to propose procedures for procurement and prepare appropriate
    tender documents (with similar contents as for water supply -
    see Paragraph 1.1.2), whereby international procurement will
    be considered

g) to prepare lists of equipment and materials, appropriate specifications and contract packages

h) to assist the Client in the evaluation of bids submitted and prepare the detailed engineering report.

1.3.3   Desludging Equipment

a) to prepare specifications for all types of equipment considered: vacuum trucks 6 m3 and 2 m3 and vacuum trailer 0.5 m3

b) to propose procedures for procurement and prepare appropriate tender documents (with similar contents as for water supply - see Paragraph 1.1.2), whereby international procurement will be considered

c) to prepare lists of equipment and materials, appropriate specifications and contract packages

h) to assist the Client in the evaluation of bids submitted and prepare the detailed engineering report.

1.4   SURFACE MICRO-DRAINS

1.4.1   The detailed engineering for the improvement of the physical status of micro-drains will comprise the following work:

a) to carry out a field investigation of the existing micro-drain system within the project area.

b) to show on maps which sections of drains should be constructed (estimate: 24 km) and those which need repairing (estimate: 177 km)

c) to describe the work to be done (type and length of drains to be constructed, kind and length of repair work, etc.) with specification of material and labor requirements, as well as estimates of time needed and cost

d) to prepare a detailed engineering design for new drains, whereby existing KIP standards and specifications could be applied

e) to propose procedures for procurement and prepare appropriate tender documents for bidding (with similar contents as for water supply - see Paragraph 1.1.2), whereby only local procurement will be considered

f) to prepare lists of materials, appropriate specifications and
    contract packages

g) to assist the Client in the evaluation of bids submitted and
    prepare the detailed engineering report.

2   SUPERVISION OF CONSTRUCTION
        _____

The scope of work during construction will comprise the
following:

a) to assist the Client in issuing instructions to the Contrac-
    tors and equipment suppliers according to the contract docu-
    ments and considering their effect on the cost of the work

b) to assist the Client in issuing instructions to Contractors
    on the extent of special inspections and testing required

c) to assist the Client in the preparation of detailed work
    schedules using up-to-date planning techniques as agreed upon
    with the Client

d) to check on the timely ordering and supply of materials and
    equipment

e) to supervise the permanent work to ensure that it is executed
    in the correct line and level and that the materials and
    workmanship comply with the specifications

f) to execute or supervise tests to be carried out on site and
    inspect materials and manufacture at source

g) to perform or supervise special inspections and testing
    necessary for the acceptance of construction works

h) to keep a diary, constituting a detailed history of the work
    in construction and all events at the site, and submit
    regular progress reports to the Client

i) to measure, in agreement with the Contractors' staff, the
    quantities of work executed, and check day work and other
    accounts in order that the interim and final payments due to
    the Contractors may be certified by the Consultants

j) to record and check the progress of the work in comparison
    with the program

k) to advise the Client of any developments threatening the delay of completion and recommend action to facilitate timely completion

l) to examine the methods proposed by the Contractors for the execution of the work and of the temporary work undertaken by them, the primary object being to ensure the safe and satisfactory execution of the permanent work

m) to review all invoices presented by the Contractors for payment, certify that the quantity, quality and cost of the materials, equipment, work performed and/or services listed therein comply with the terms of the Construction Contract

n) to review and evaluate claims for extra payment by the Contractors and equipment suppliers, and make recommendations thereon to the Client, as well as assist in resolving disputes arising therefrom

o) to assist the Client in commissioning and handing over of the completed works to the responsible authority after final inspection and issue the final certificate, ensure the completion and handing over of record drawings

p) to prepare progress charts or diagrams

q) to issue monthly progress reports

r) to proposed remedial measures if progress is behind schedule.

3   PACKAGING AND SCHEDULING

The packaging and scheduling for the implementation of the project should be based on the following principles:

a) to avoid long-term disturbance to the population in kampungs due to the construction work in the streets, all sanitation facilities in a kampung should be built, to the maximum possible extent, in the same period; this particularly concerns the construction of the piped water distribution network with public water taps, repair and construction of drains

b) to prepare the construction packages per kampung or group of
    kampungs, depending on the size of kampungs or the volume of
    the work to be implemented

c) to respect particular requirements or procedures for the work
    to be implemented, e.g. deep well stations (requiring approv-
    al from the Ministry of Mines for drilling, PLN for electri-
    city supply, etc.), desludging equipment, transfer and thick-
    ening station (international bidding)

d) the rehabilitation/repair of facilities (public water taps,
    MCK, drains) to be given priority.

Except for the transfer and thickening station, which shall be
built within the construction package of water treatment plant
proposed at Setia Budi ponds, the packaging and scheduling prac-
ticed by DKI/KIP could be applied. Accordingly, the amount involved would be from about Rp. 5.0 million (US $ 7,700) to Rp.
50.0 million (US $77,000) per contractor depending on the capa-
city, experience and ability of prequalified contractors.[1]

The grouping of the Construction Contracts could be conveniently
done for extension of the piped water distribution network,
repair and construction of drains and the construction of public
water taps, MCKs and leaching pits. Other work such as the rehabilitation of existing public water taps and MCKs as well as
desludging equipment and the transfer and thickening station
could easily be done in separate packages.

4   PROCUREMENT
        _____

Procurement procedure should be consistent with the World Bank
guidelines where international competitive bidding has to be
considered, such as for desludging equipment and the transfer and
thickening station; for the remaining sanitation components,
where only local competitive bidding is to be organized,

---

1) There were 496 small local contractors registered and prequalified by DKI/KIP for implementation of the KIP project in
   1982.

Accordingly, the procurement procedures shall include:

(i) preparation of tender documents comprising all relevant data and information concerning the project, with appropriate drawings, technical specifications and instructions for bidding

(ii) a reasonable period for preparing bids and quotations to be submitted by contractors and/or suppliers, whereby all contractors/suppliers should be appropriately prequalified as to their ability and capacity to execute the work

(iii) a thorough evaluation of bids submitted, whereby the following criteria have to be taken into consideration for the selection of the contractor/supplier:

- expected quality of work
- the cost of work and services offered
- schedule of implementation
- conditions of payment
- guarantee

(iv) the definite selection of contractors/suppliers should be approved by a Tendering Board consisting of delegates/ representatives of all government agencies involved (DKI/KIP, PAM, DK[1]) and Public Works)

(v) award of contracts.

Generally, the local procurement procedures practised by DKI/KIP could be applied.

5   TRAINING
        ─────────

With the aim of transferring skill and knowledge concerning planning, establishment, operation, maintenance and monitoring of all sanitation components concerned, an appropriate program of training local staff will be set up and executed during the implementation of the project.

This program will include:

(i) theoretical training concerning specific aspects of the sanitation project, including all individual sanitation

─────────────────────
1) DK = Dinas Kebersihan (Cleaning Department)

components: MCKs, water supply, leaching pits, desludging services and surface drains

(ii)   practical on-the-job training of technical and administrative staff engaged for the above activities

(iii)   detailed time schedules of training, labor, equipment and material requirements, cost estimates and financing.

6   PUBLIC RELATIONS

To ensure the expected efficiency of the proposed sanitation improvements, consulting services will include the following:

a)   in close cooperation with DKI and through DKI with other agencies concerned, to assist the Client in the preparation of an educational campaign of the population in the area with the aim of obtaining their support for and acceptance of the new sanitation facilities (to use MCKs instead of using rivers for bathing, washing and as lavatories, to provide land for MCKs, deep well stations and public water taps, to build leaching pits for their toilets, to participate in cleaning/ maintenance of MCKs, drains and ditches, etc.) as well as to inform them of hygienic principles in using water, toilets and other sanitation services. In this connection, a monitoring program for all sanitation facilities could be used as well

b)   to provide the assistance of a sociologist (at least on a part-time basis) during the above campaing

c)   close cooperation with all local political and religious leaders in the area

d)   to evaluate the progress of the campaign and propose appropriate measures for its further development.

7 OPERATION, MAINTENANCE AND MONITORING

To ensure that the rehabilitated/reactivated and newly established sanitation facilities will operate appropriately, an adequate operation, maintenance and monitoring system covering all the above-mentioned facilities shall be set up. For this purpose, engineering services will include:

7.1 OPERATION AND MAINTENANCE

a) to carry out a detailed analysis of existing operation and maintenance systems for sanitation facilities and propose measures for their improvement, whereby particular attention shall be paid to desludging services, MCKs and surface micro-drains

b) in this connection, the proposals will determine, for each particular sanitation component:

 (i) what agency and who within the agency is responsible for O & M of the facility

 (ii) how the facility should be properly operated and maintained

 (iii) who should supervise O & M and how often this supervision must be executed

 (iv) staff and budget requirements

 (v) how the people themselves could/must contribute to O & M of the facilities they use as well as to the covering of financial costs: e.g. what fees could be required from the users of a facility and if no payment is recommended (for poor families) how the O & M cost could/should be covered

 (vi) job descriptions for operators/managers/supervisors, including education and experience requirements.

7.2 MONITORING

a) to analyze the existing monitoring system and actual monitoring activities for each sanitation component

b) to propose improvements of the monitoring system covering both (i) sanitation facilities in the project area including "old" KIP as well as the new ones which should be operated by Dinas Kebersihan, PAM and DSS[1] and (ii) sewerage facilities

c) in this connection, the proposals will determine, for all the above facilities:

   (i)   how often to monitor

   (ii)   how often to report

   (iii)   who should receive reports (government agency and official responsible)

   (iv)   need for coordinating committees (e.g. DSS and Dinas Kebersihan for solid waste and desludging, DSS and PAM for water supply, etc.)

   (v)   monitoring organization and flow chart system of information to be submitted to the government agency responsible for O & M of the sanitation facility concerned

   (vi)   formats for inspectors to use

   (vii)   formats for periodic reports

   (viii) staff and budget requirements to make the monitoring program operational

   (ix)   job description for inspectors, including education and experience requirements

   (x)   description of the information to be collected routinely by RT and/or RW on both physical and socio-economic parameters, for example, how much water is sold, number of families using it, how much is consumed per family, amount paid to PAM, role of vendors, are the poor people getting service, etc. Also, for example, for MCK, is it accepted and used as planned to achieve its purpose, is it used, is it clean, who manages it, how is the manager selected and paid, how much are the fees, is it preventing "going to the fields and rivers", what needs to be done to make up for deficiencies, etc.

d) monitoring should include both the need for physical repairs and socioeconomic aspects, such as:

   (i)   who the users are, how they are served by the facility and how much they pay for services (by discussion with holder/operator/manager and by observation)

---

[1] DSS = Division of Sewerage and Sanitation

(ii)   amount paid/transferred for services (water, de-sludging, use of MCKs) to appropriate government agency involved (by discussion with holder/operator/manager and checking of agency's records)

(iii)   extent to which the low-income population of the kampung (the intended beneficiaries) are being adequately served, or not served and why not

(iv)   adequacy of services from the customer point of view as to price/quantity/quality, especially with respect to the low-income population

(v)   the engineering history of the new facility, including construction period, date of commissioning, construction cost, problems, how the land was obtained, its ownership and how this ownership relates to the present holder

(vi)   the history of the holder/operator/manager of the facility: how was he selected and his relationship/agreements with the local RT, RW and Lurah, including financial agreements

e)   for monitoring the project, appropriate performance indicators for each sanitation component shall be set up and regularly reviewed each year; for illustration, see the <u>attached</u> list of such performance indicators for water supply.

Performance Indicators for Monitoring
Water Supply
(Example)

	YEAR			
	1983	1984	1985	1986

1. <u>Water Production</u> (1.00 m3)

   a) PAM - mains
   b) deep well stations

   Total

2. <u>Water Sold</u> (1.000 m3)

   a) metered house connections
   b) standpipe on plots
   c) Public water taps/cisterns:
      i) direct at water taps
      ii) through vendors

   Total

   % of production

3. <u>Number of House Connections</u>

   Number of standpipes on plot
   Number of public water taps

4. <u>Population Served</u>

   a) by house connections
   b) by standpipes on plot
   c) by public water taps

5. <u>Water Consumption</u> (lpcd)

   a) through house connection
   b) through standpipe on plot
   c) through public water tap

6. <u>Water Sales</u> (Million Rp.)

   a) house connections
   b) standpipes on plot
   c) public water taps

ANNEX 29
SANITATION SURVEY: Required Staff, Scheduling and Transportation

	Unit	25	50	100	200	300	400	500	600	700	800	900	1,000
Total Population of the Area	1,000	25	50	100	200	300	400	500	600	700	800	900	1,000
Population to be Surveyed	%	10.0	9.0	8.0	6.0	5.0	4.5	4.2	4.0	3.7	3.5	3.2	3.0
	1,000	2.5	4.5	8	12	15	18	21	24	26	28	29	30
Total Field Working Days[1]	man/day	84	150	267	400	500	600	700	800	867	933	967	1,000
Professional Staff:													
– Study team leader	man	1	1	1	1	1	1	1	1	1	1	1	1
	man/day	63	63	63	73	84	84	94	94	105	105	105	105
– Sanitary engineer(s)	man	1	1	1	1	1	1	1	2	2	2	2	2
	man/day	63	63	63	73	84	84	84	115	130	130	130	130
– Surveyors[2]	man	4	8	14	14	14	16	18	20	22	24	24	26
	man/day	92	176	308	448	546	640	756	860	946	1,008	1,032	1,066
– Coordinator	man	–	–	1	1	1	1	1	1	2	2	2	2
	man/day	–	–	22	32	39	40	42	43	86	84	86	82
Supporting Staff:													
– Draftsman (–men)	man	1	1	1	1	1	1	1	2	2	2	2	2
	man/day	63	63	63	73	84	84	94	115	130	130	130	130
– Secretary	man	1	1	1	1	1	1	1	1	1	1	1	1
	man/day	63	63	63	73	84	84	94	94	105	105	105	105
TOTAL	man	8	12	19	19	19	21	23	27	30	32	32	34
	man/day	344	428	582	772	921	1,016	1,164	1,321	1,502	1,562	1,588	1,618
Field work	day[3]	23	22	22	32	39	40	42	43	43	42	43	41
	month[4]	1	1	1	1.5	2	2	2	2	2	2	2	2
Total study	day	63	63	63	73	84	84	94	94	105	105	105	105
	month	3	3	3	3.5	4	4	4.5	4.5	5	5	5	5
Private car(s)	number	1	1	2	2	2	2	2	3	3	3	3	3
Mini–bus(es)	number	1	1	2	2	2	2	2	3	3	3	3	3
TOTAL	number	2	2	4	4	4	4	4	6	6	6	6	6

(Side groupings: Staffing / Schedule / Transport)

Notes: 1) Based on ratio: 30 persons/surveyor/working day
2) Field working days + training days
3) Based on estimates of working groups (see Annex 30 + training of surveyors (2 – 3 days)
4) Month = 30 calendar days (21 working days)

ANNEX 30

SANITATION SURVEY: Estimates of Field Working Days (FWD) Depending on Population to- be Surveyed

Number of Surveyors	Number of Groups (1)	FWD per 1,000 Persons (2)	Field working days for population to be surveyed – see Annex 46											
			2,500	4,500	8,000	12,000	15,000	18,000	21,000	24,000	26,000	28,000	29,000	30,000
2	1	16.66	42	75	133	200	250	300	350	400	433	466	483	500
4	2	8.33	21	38	67	100	125	150	175	200	217	233	242	250
6	3	5.55	14	25	44	67	83	100	117	133	144	155	161	167
8	4	4.17	10	19	33	50	63	75	88	100	108	117	121	125
10	5	3.33	8	15	27	40	50	60	70	80	87	93	97	100
12	6	2.78	7	13	22	33	42	50	58	67	72	78	81	83
14	7	2.38		11	19	29	36	43	50	57	62	67	69	71
16	8	2.08		9	17	25	31	37	44	50	54	58	60	62
18	9	1.85		8	15	22	28	33	39	44	48	52	54	56
20	10	1.67		7	13	20	25	30	35	40	43	47	48	50
22	11	1.52			12	18	23	27	32	36	40	43	44	46
24	12	1.39			11	17	21	25	29	33	36	39	40	42
26	13	1.28			10	15	19	23	27	31	33	36	37	38
28	14	1.19			9	14	18	21	25	29	31	33	35	36
30	15	1.11			9	13	17	20	23	27	29	31	32	33
32	16	1.04			8	12	16	19	22	25	27	29	30	31
34	17	0.98			8	12	15	18	21	24	25	27	28	29
36	18	0.93			7	11	14	17	20	22	24	26	27	28
38	19	0.88				11	13	16	18	21	23	25	26	26
40	20	0.83				10	12	15	17	20	22	23	24	25
42	21	0.79				9	12	14	17	19	21	22	23	24
44	22	0.76				9	11	14	16	18	20	21	22	23
46	23	0.72				9	11	13	15	17	19	20	21	22
48	24	0.69				8	10	12	14	17	18	19	20	21
50	25	0.67				8	10	12	14	16	17	18	19	20

(1) Two surveyors per group
(2) Based on ratio 11 households to be interviewed (i.e. about 60 persons to be surveyed) per group/day

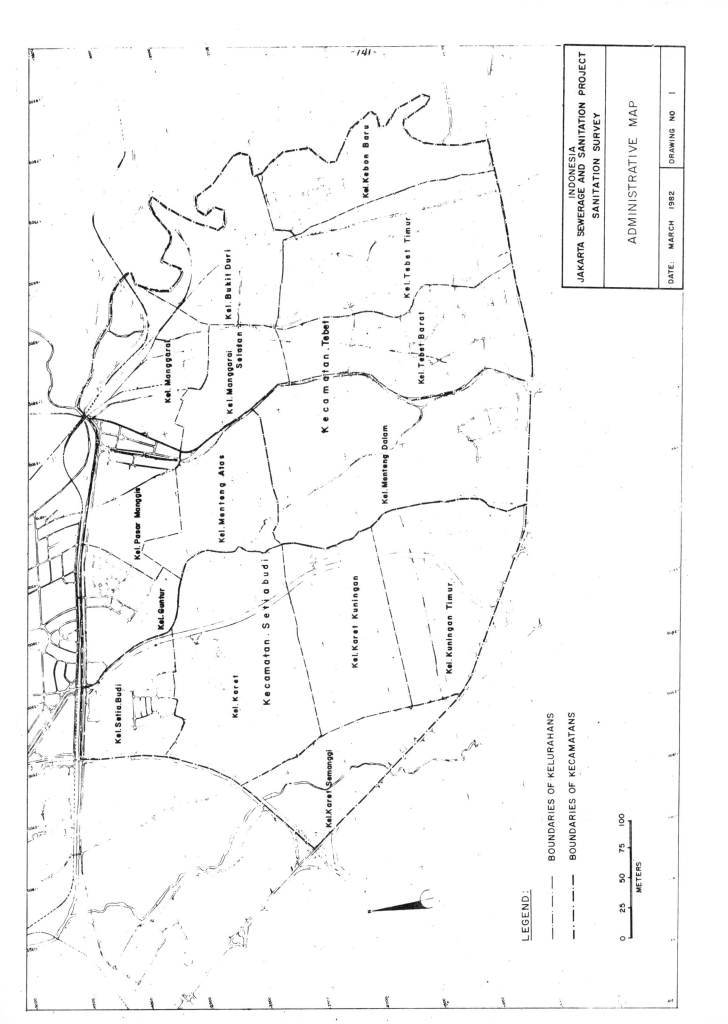

INDONESIA
JAKARTA SEWERAGE AND SANITATION PROJECT
SANITATION SURVEY

ADMINISTRATIVE MAP

DATE: MARCH 1982   DRAWING NO 1

Kel.Kebon Baru

Kel.Tebet Timur

Kel.Bukit Duri

Kel.Manggarai
Selatan

Kel.Manggarai

Kecamatan.Tebet

Kel.Tebet Barat

Kel.Menteng Atas

Kel.Menteng Dalam

Kel.Pasar Manggis

Kel.Guntur

Kecamatan.Setiabudi

Kel.Karet Kuningan

Kel.Setia.Budi

Kel.Karet

Kel.Kuningan Timur

Kel.Karet Semanggi

LEGEND:

— · — · —   BOUNDARIES OF KELURAHANS

— · — ·   BOUNDARIES OF KECAMATANS

0   25   50   75   100
METERS

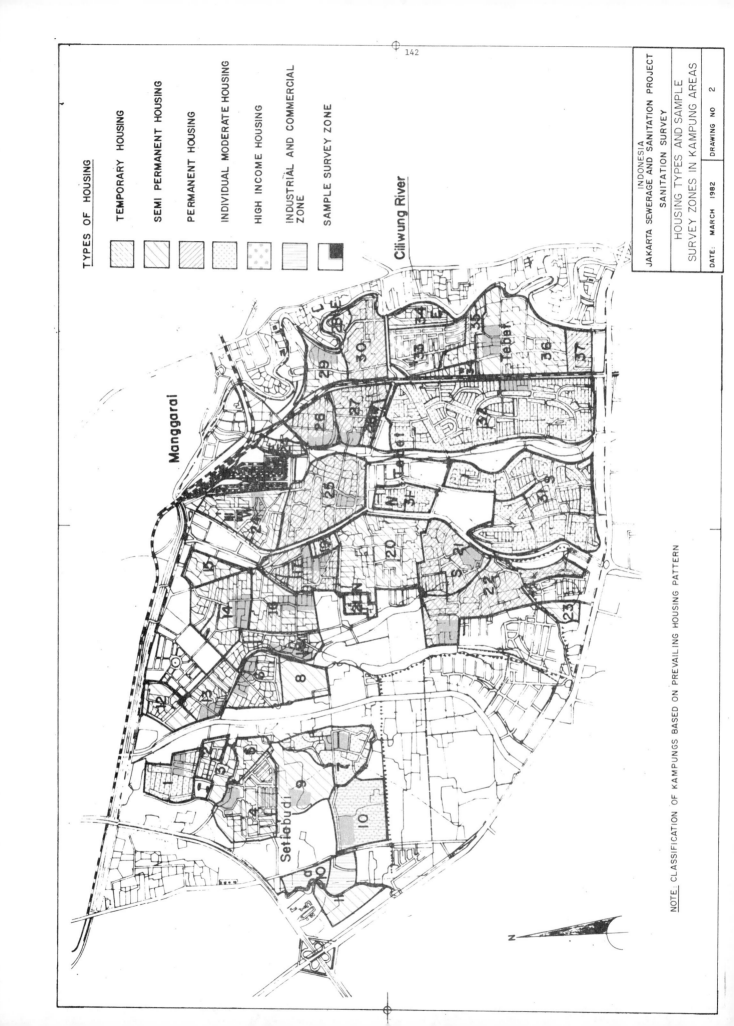

TYPES OF HOUSING

TEMPORARY HOUSING

SEMI PERMANENT HOUSING

PERMANENT HOUSING

INDIVIDUAL MODERATE HOUSING

HIGH INCOME HOUSING

INDUSTRIAL AND COMMERCIAL ZONE

SAMPLE SURVEY ZONE

Ciliwung River

Manggarai

Tebet

Setiabudi

INDONESIA
JAKARTA SEWERAGE AND SANITATION PROJECT
SANITATION SURVEY

HOUSING TYPES AND SAMPLE
SURVEY ZONES IN KAMPUNG AREAS

DATE: MARCH 1982   DRAWING NO 2

NOTE: CLASSIFICATION OF KAMPUNGS BASED ON PREVAILING HOUSING PATTERN

N

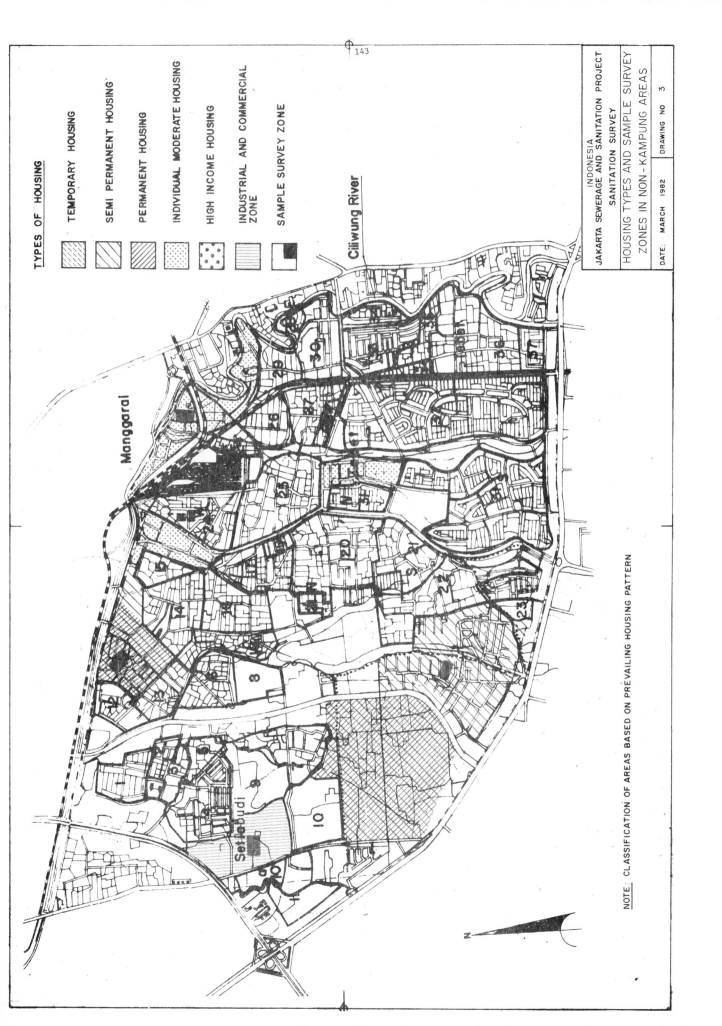

TYPES OF HOUSING

TEMPORARY HOUSING

SEMI PERMANENT HOUSING

PERMANENT HOUSING

INDIVIDUAL MODERATE HOUSING

HIGH INCOME HOUSING

INDUSTRIAL AND COMMERCIAL ZONE

SAMPLE SURVEY ZONE

Ciliwung River

Manggarai

Setiabudi

N

INDONESIA
JAKARTA SEWERAGE AND SANITATION PROJECT
SANITATION SURVEY

HOUSING TYPES AND SAMPLE SURVEY
ZONES IN NON-KAMPUNG AREAS

DATE: MARCH 1982   DRAWING NO 3

NOTE: CLASSIFICATION OF AREAS BASED ON PREVAILING HOUSING PATTERN

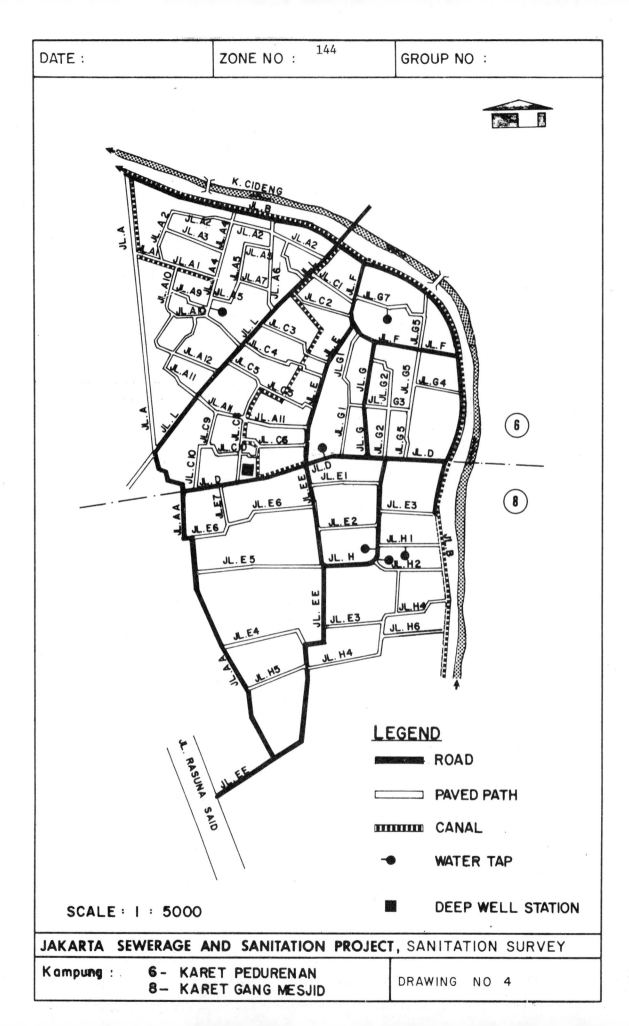

DATE :

ZONE NO : 144

GROUP NO :

K. CIDENG

6

8

LEGEND

▬▬ ROAD

▭ PAVED PATH

▥▥ CANAL

●─ WATER TAP

■ DEEP WELL STATION

SCALE : 1 : 5000

JL. RASUNA SAID

JAKARTA SEWERAGE AND SANITATION PROJECT, SANITATION SURVEY

Kampung : 6 - KARET PEDURENAN
8 - KARET GANG MESJID

DRAWING NO 4

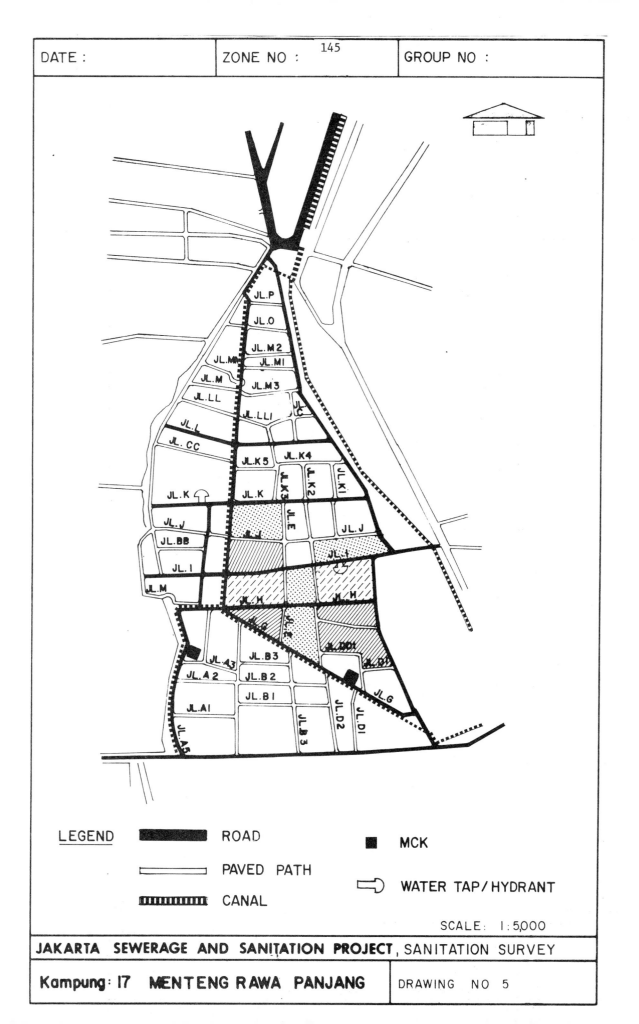

| DATE : | ZONE NO : 145 | GROUP NO : |

LEGEND
ROAD
PAVED PATH
CANAL
MCK
WATER TAP / HYDRANT

SCALE: 1:5,000

JAKARTA SEWERAGE AND SANITATION PROJECT, SANITATION SURVEY

Kampung: 17 MENTENG RAWA PANJANG | DRAWING NO 5

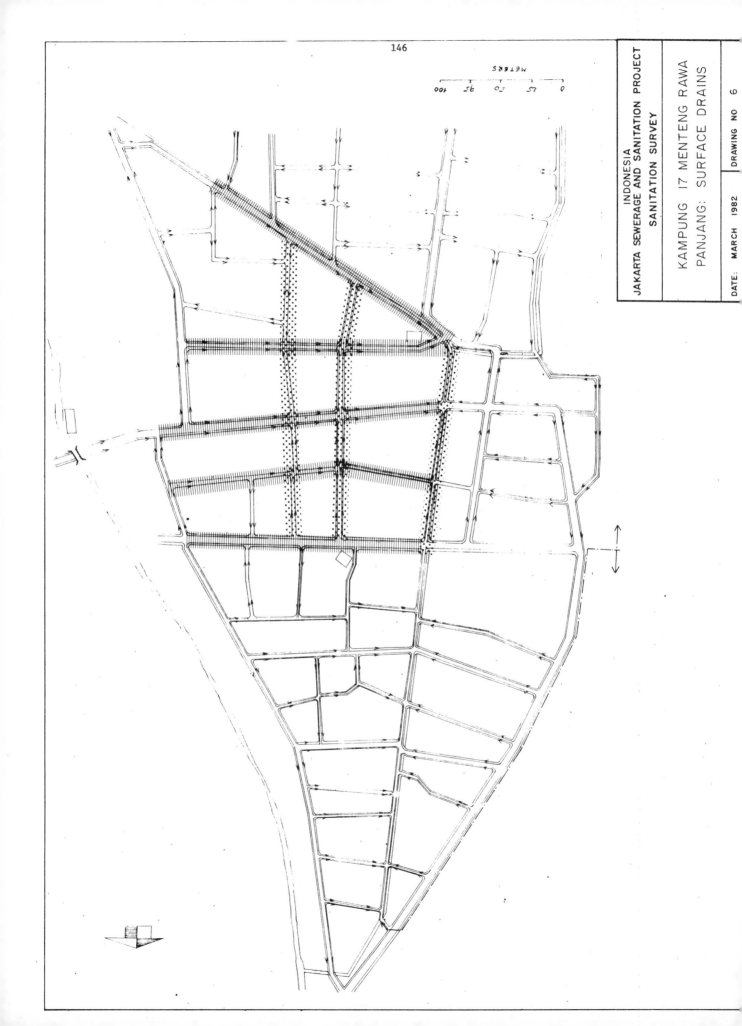

INDONESIA
JAKARTA SEWERAGE AND SANITATION PROJECT
SANITATION SURVEY

KAMPUNG 17 MENTENG RAWA
PANJANG: SURFACE DRAINS

DATE: MARCH 1982   DRAWING NO 6

METERS

0 25 50 75 100

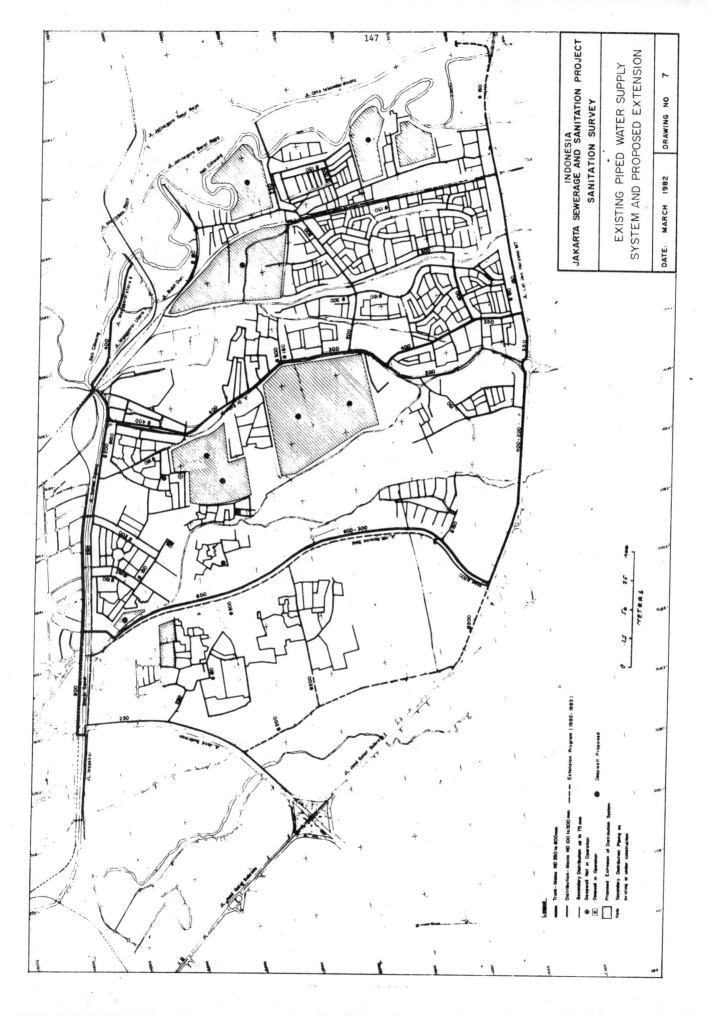

INDONESIA
JAKARTA SEWERAGE AND SANITATION PROJECT
SANITATION SURVEY

EXISTING PIPED WATER SUPPLY
SYSTEM AND PROPOSED EXTENSION

DATE: MARCH 1982   DRAWING NO 7

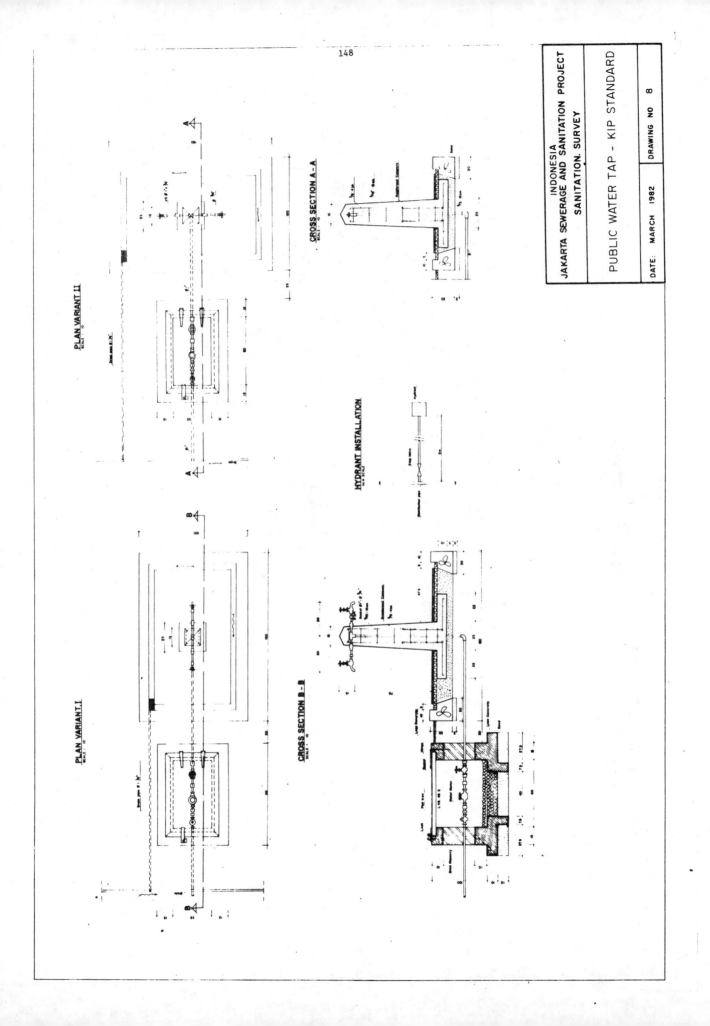

PLAN VARIANT II

PLAN VARIANT I

CROSS SECTION A-A

CROSS SECTION B-B

HYDRANT INSTALLATION

INDONESIA
JAKARTA SEWERAGE AND SANITATION PROJECT
SANITATION. SURVEY

PUBLIC WATER TAP - KIP STANDARD

DATE: MARCH 1982   DRAWING NO 8

149

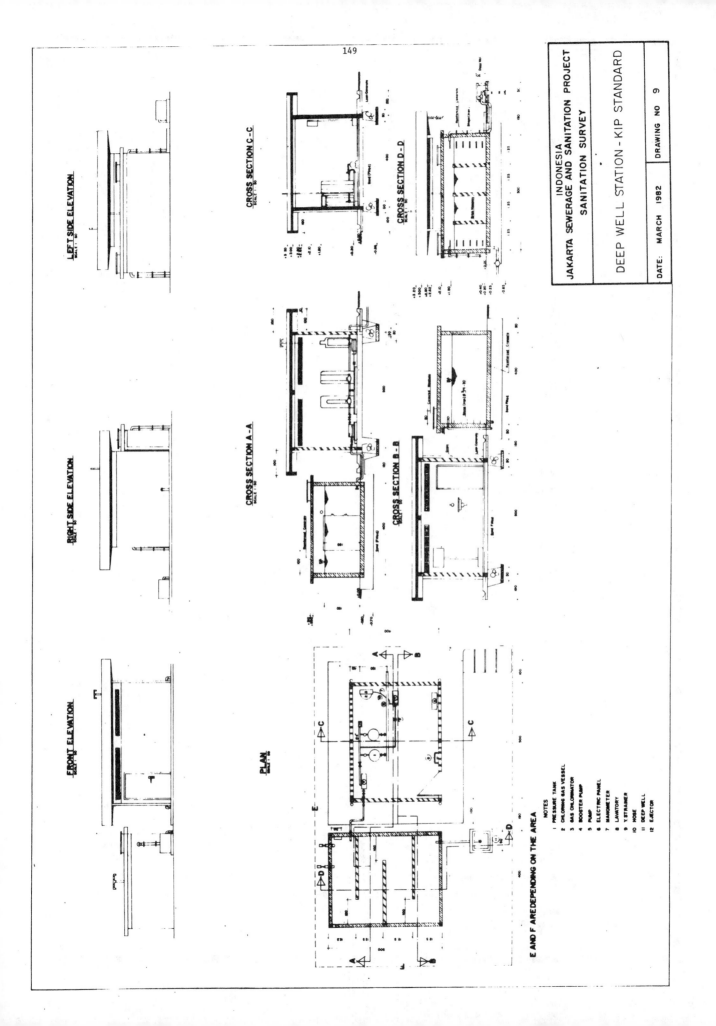

LEFT SIDE ELEVATION

RIGHT SIDE ELEVATION

FRONT ELEVATION

CROSS SECTION C-C

CROSS SECTION A-A

PLAN

CROSS SECTION D-D

CROSS SECTION B-B

NOTES
1 PRESSURE TANK
2 CHLORINE GAS VESSEL
3 GAS CHLORINATOR
4 BOOSTER PUMP
5 PUMP
6 ELECTRIC PANEL
7 MANOMETER
8 LAVATORY
9 Y-STRAINER
10 HOSE
11 DEEP WELL
12 EJECTOR

E AND F ARE DEPENDING ON THE AREA

INDONESIA
JAKARTA SEWERAGE AND SANITATION PROJECT
SANITATION SURVEY

DEEP WELL STATION - KIP STANDARD

DATE: MARCH 1982   DRAWING NO 9

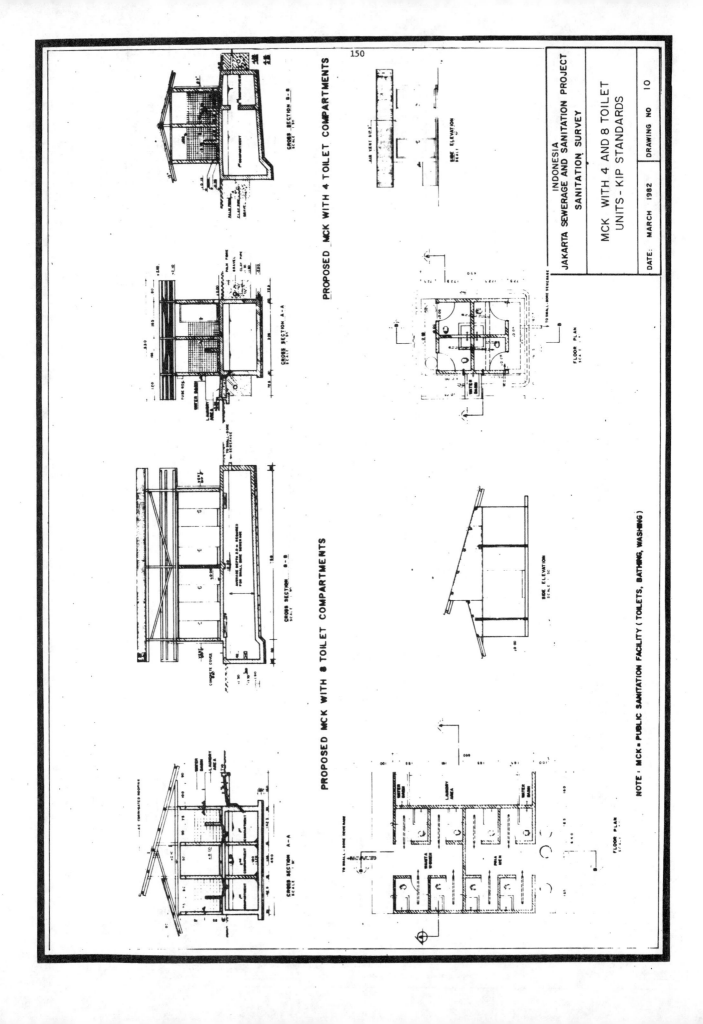

PROPOSED MCK WITH 4 TOILET COMPARTMENTS

PROPOSED MCK WITH 8 TOILET COMPARTMENTS

NOTE : MCK = PUBLIC SANITATION FACILITY (TOILETS, BATHING, WASHING)

INDONESIA
JAKARTA SEWERAGE AND SANITATION PROJECT
SANITATION SURVEY

MCK WITH 4 AND 8 TOILET
UNITS - KIP STANDARDS

DATE: MARCH 1982   DRAWING NO   10

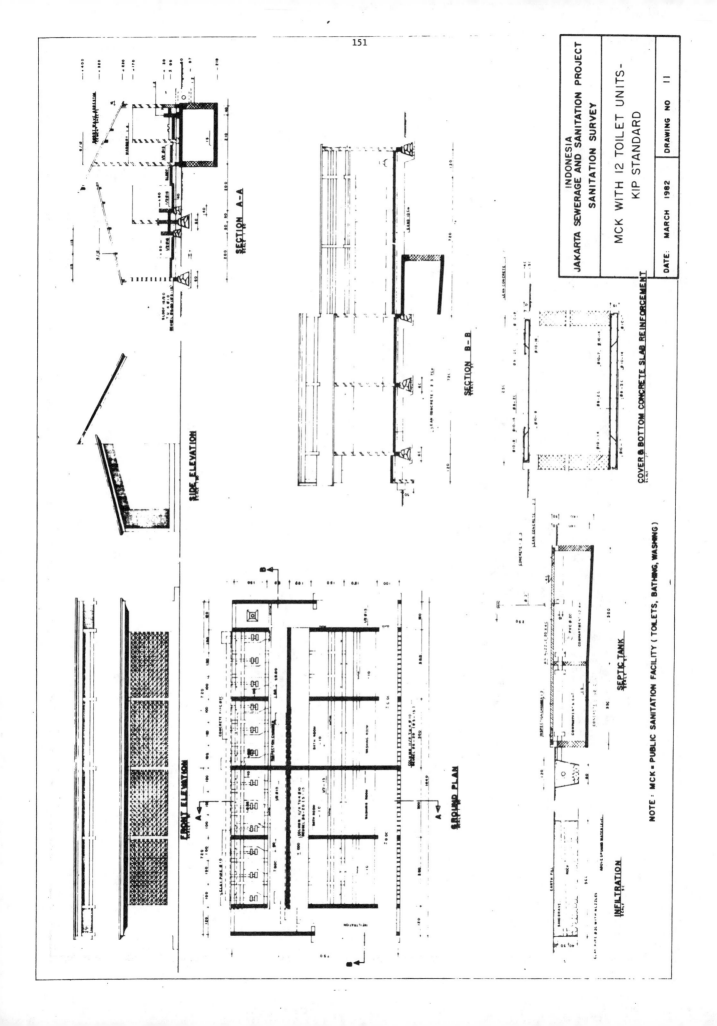

INDONESIA
JAKARTA SEWERAGE AND SANITATION PROJECT
SANITATION SURVEY

MCK WITH 12 TOILET UNITS-
KIP STANDARD

DATE: MARCH 1982   DRAWING NO 11

SECTION A-A

SIDE ELEVATION

SECTION B-B

COVER & BOTTOM CONCRETE SLAB REINFORCEMENT

FRONT ELEVATION

GROUND PLAN

SEPTIC TANK

INFILTRATION

NOTE : MCK = PUBLIC SANITATION FACILITY (TOILETS, BATHING, WASHING)

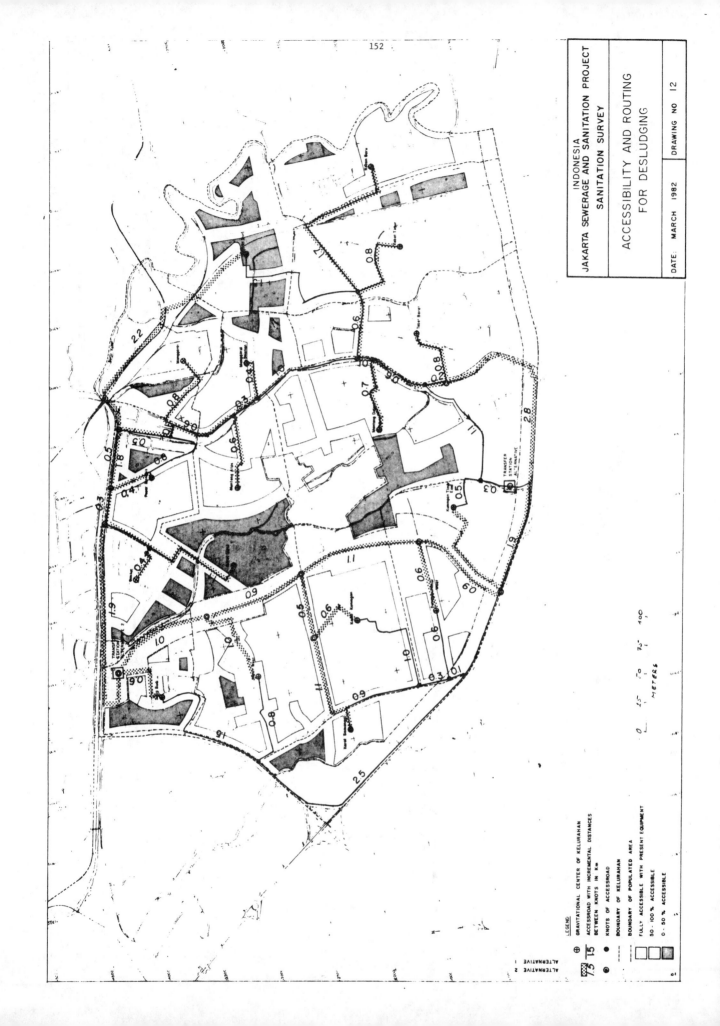

INDONESIA
JAKARTA SEWERAGE AND SANITATION PROJECT
SANITATION SURVEY

ACCESSIBILITY AND ROUTING
FOR DESLUDGING

DATE: MARCH 1982   DRAWING NO 12

LEGEND

GRAVITATIONAL CENTER OF KELURAHAN

ACCESSROAD WITH INCREMENTAL DISTANCES
BETWEEN KNOTS IN Km

KNOTS OF ACCESSROAD

BOUNDARY OF KELURAHAN

BOUNDARY OF POPULATED AREA

FULLY ACCESSIBLE WITH PRESENT EQUIPMENT

50 - 100 % ACCESSIBLE

0 - 50 % ACCESSIBLE

ALTERNATIVE 1
ALTERNATIVE 2

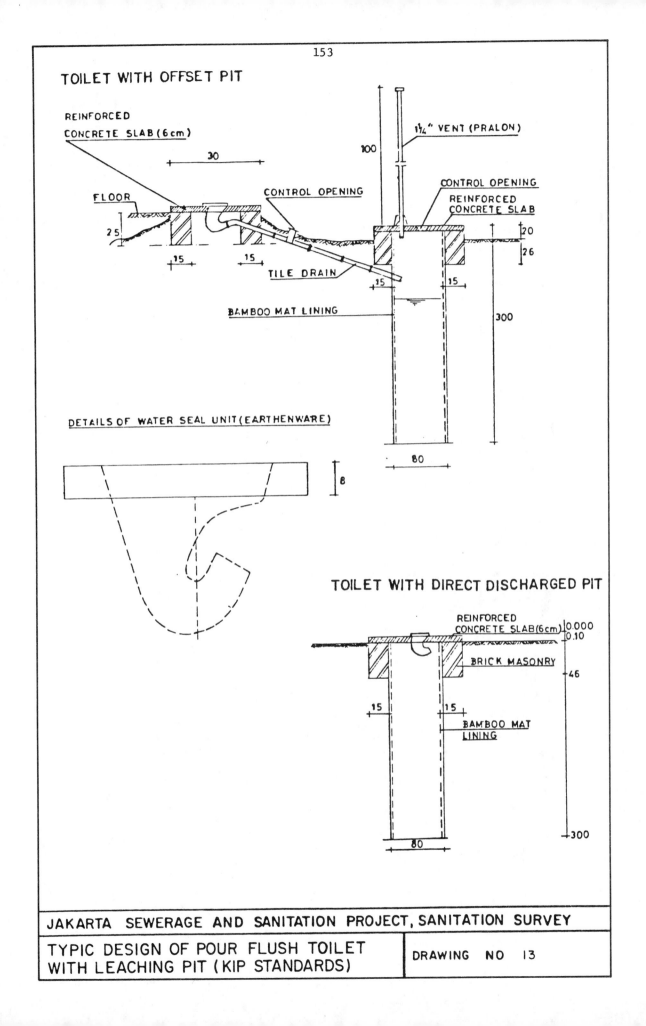

TOILET WITH OFFSET PIT

REINFORCED
CONCRETE SLAB (6 cm)

1¼" VENT (PRALON)

FLOOR

30

100

CONTROL OPENING

CONTROL OPENING
REINFORCED
CONCRETE SLAB

25

15

15

20

26

TILE DRAIN

BAMBOO MAT LINING

300

DETAILS OF WATER SEAL UNIT (EARTHENWARE)

8

80

TOILET WITH DIRECT DISCHARGED PIT

REINFORCED
CONCRETE SLAB (6 cm)

0.000
0.10

BRICK MASONRY

46

15

15

BAMBOO MAT
LINING

300

80

JAKARTA SEWERAGE AND SANITATION PROJECT, SANITATION SURVEY

TYPIC DESIGN OF POUR FLUSH TOILET
WITH LEACHING PIT (KIP STANDARDS)

DRAWING NO 13

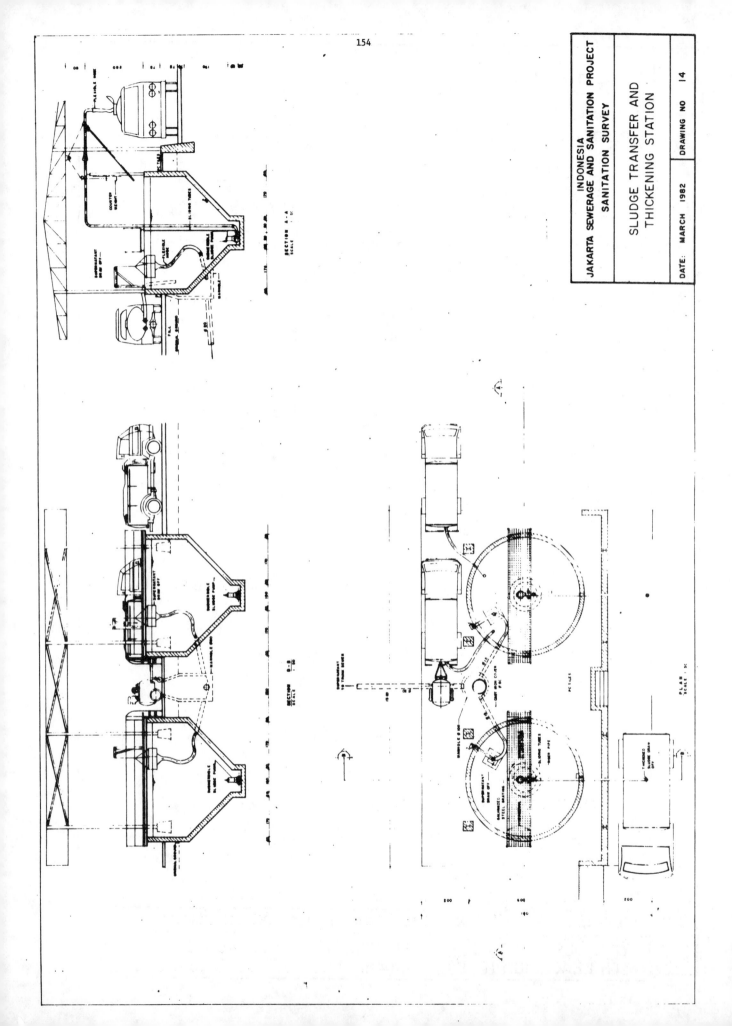

INDONESIA
JAKARTA SEWERAGE AND SANITATION PROJECT
SANITATION SURVEY

SLUDGE TRANSFER AND
THICKENING STATION

DRAWING NO 14

DATE: MARCH 1982

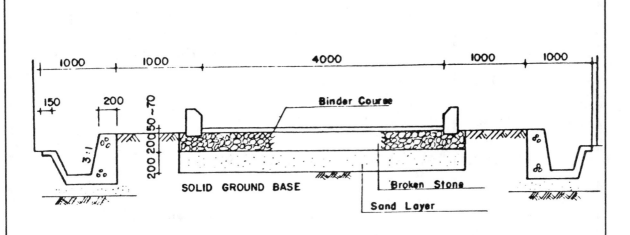

1000   1000   4000   1000   1000

150   200

Binder Course

SOLID GROUND BASE

Broken Stone

Sand Layer

CROSS SECTION OF TYPICAL VEHICULAR ROAD

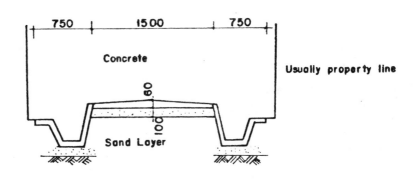

750   1500   750

Concrete

Usually property line

Sand Layer

CROSS SECTION OF TYPICAL FOOTPATH

JAKARTA SEWERAGE AND SANITATION PROJECT, SANITATION SURVEY

CROSS SECTION OF TYPICAL ROAD AND
FOOTPATH (KIP STANDARDS)

DRAWING NO 15

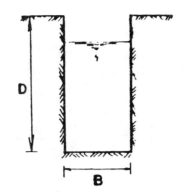

TYPE A

Rectangular Channel

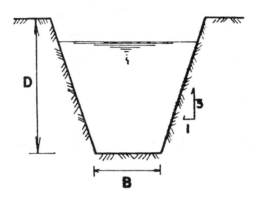

TYPE B

Trapezoidal Channel
Side Slopes 3 : 1

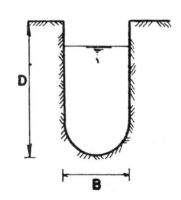

TYPE C

U — Channel

Bottom : Half of a
circular conduit

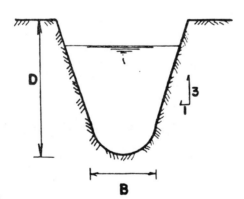

TYPE D

Trapezoidal U—Channel

Bottom : Half of a
circular conduit

Side Slopes 3 : 1

JAKARTA SEWERAGE AND SANITATION PROJECT, SANITATION SURVEY

TYPICAL DRAINAGE PROFILES
(KIP STANDARDS)

DRAWING   NO 16

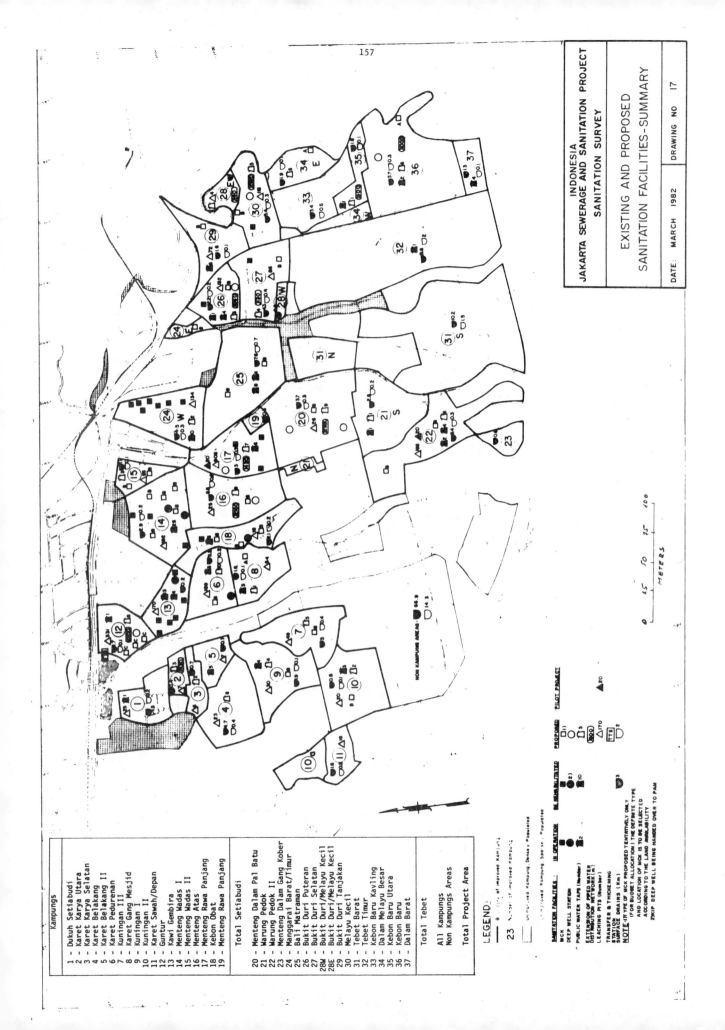

LEGEND:

Kampungs

1 - Dukuh Setiabudi
2 - Karet Karya Utara
3 - Karet Karya Selatan
4 - Karet Belakang
5 - Karet Belakang II
6 - Karet Pedurenan
7 - Kuningan III
8 - Karet Gang Mesjid
9 - Kuningan I
10 - Kuningan II
11 - Karet Sawah/Depan
12 - Guntur
13 - Kawi Gembira
14 - Menteng Wadas I
15 - Menteng Wadas II
16 - Menteng Atas
17 - Menteng Rawa Panjang
18 - Kebon Obat
19 - Menteng Rawa Panjang

Total Setiabudi

20 - Menteng Dalam Pal Batu
21 - Warung Pedok
22 - Warung Pedok II
23 - Menteng Dalam Gang Kober
24 - Manggarai Barat/Timur
25 - Bali Matraman
26 - Bukit Duri Puteran
27 - Bukit Duri Selatan
28W - Bukit Duri/Melayu Kecil
28E - Bukit Duri/Melayu Kecil
29 - Bukit Duri Tanjakan
30 - Melayu Kecil
31 - Tebet Barat
32 - Tebet Timur
33 - Kebon Baru Kaviing
34 - Dalam Melayu Besar
35 - Kebon Baru Utara
36 - Kebon Baru
37 - Dalam Barat

Total Tebet

All Kampungs
Non Kampungs Areas

Total Project Area

SANITATION FACILITIES:
MCK
DEEP WELL STATION
PUBLIC WATER TAPS (Number)
EXTENSION OF PIPED WATER
DISTRIBUTION NETWORK (m)
LEACHING PITS (Number)
TRANSFER & THICKENING
STATION
SURFACE DRAINS (Km)
NOTE (IF TYPE OF MCK PROPOSED TENTATIVELY ONLY
AND BUDGET ALLOCATION) THE DEFINITE TYPE
AND LOCATION OF MCK IS TO BE SELECTED
ACCORDING TO THE LAND AVAILABILITY
ZMIP DEEP WELL BEING HANDED OVER TO PAM

IN OPERATION   RE-HABILITATED   PROPOSED   PILOT PROJECT

INDONESIA
JAKARTA SEWERAGE AND SANITATION PROJECT
SANITATION SURVEY

EXISTING AND PROPOSED
SANITATION FACILITIES-SUMMARY

DATE: MARCH 1982   DRAWING NO 17

BIBLIOGRAPHY

(List of References)

"Jakarta Sewerage and Sanitation Project", series of volumes prepared by
    Alpinconsult/P.T. Encona for Directorate General of Cipta Karya,
    May and September 1982, including:
    (1/1) "Volume 1, Main Report".
    (1/8) "Volume 8, Sanitation Study".
    (1/9) "Volume 9, Sanitation Study Drawings".

"Immediate Programme of Sanitation", by H.F. Ludwig/Nihon Suido, one of
    series of volumes making up the report, "Comprehensive Plan of
    Sewerage and Sanitation for Metropolitan Jakarta", prepared by Nihon
    Suido for UNDP/WHO/Directorate General of Cipta Karya, 1977.

"Report on Maintenance Aspects of Indonesia KIP Programs at Jakarta and
    Surabaya for Roads, Footpaths, Drains, Water Taps, and MCKs",
    by H.F. Ludwig for World Bank, April 1980.

"Report on IBRD Supervisory Mission for Kampung Improvement Projects of
    Indonesia, Urban II/III Water Supply for Jakarta and Surabaya",
    by H.F. Ludwig for World Bank, Oct. 1979.

"Quantification of Impact of Water and Sanitation Improvements in Urban Low
    Income Communities at Jakarta", H.F. Ludwig, UNADI/UNEP Seminar on
    Environment, Bangkok, 1976.

"Appropriate Technology for Water Supply and Sanitation, Volume 2, a Planner's
    Guide", by J.M. Kalbermatten et al, World Bank, Dec. 1980.

"Appropriate Technology for Water Supply and Sanitation, Volume 3, Health
    Aspects of Excreta and Sullage Management, a State-of-the-Art Review",
    by R.G. Feachem et al, World Bank, 1980 .

"Appropriate Technology for Water Supply and Sanitation, Volume 5, Socio-
    cultural Aspects of Water Supply and Excreta Disposal", by M. Elimendorf
    and P. Buckles, World Bank, 1980.

"Appropriate Technology for Water Supply and Sanitation, Volume 11, A
    Sanitation Field Manual, by J.M. Kalbermatten et al, World Bank,
    1980.

"Urban Sewerage and Excreta Disposal Planning for Chonburi and Thailand",
    (Draft Report), by Seatec International for WHO/Department of Public
    Works, Bangkok, Jan. 1983.

"An Integrated Monitoring and Evaluation Study of the Kampung Improvement
    Programs in Five Indonesian Cities", by Institute for Economic and
    Social Research, Education, and Information, Jakarta, 1980.

World Bank Publications of Related Interest

Laboratory Evaluation of Hand-operated Pumps for Use in Developing Countries
Consumers' Association Testing and Research Laboratory

An invaluable guide for selecting water pumps. Paper rates 12 pumps on basis of rigorous testing to determine their strengths and weaknesses. Pumps rated for packaging, condition as received, installation, maintenance, repair, and overall desirability.

Technical Paper No. 6. 1983. 89 pages.

ISBN 0-8213-0133-0. Stock No. BK 0133. $3.

Laboratory Testing, Field Trials, and Technological Development

Contains results of laboratory tests carried out on twelve hand pumps for the World Bank and presents recommendations for improvements in performance, safety, and durability.

1982. 122 pages (including 3 appendixes).

Stock No. BK 9190. $5.

Meeting the Needs of the Poor for Water Supply and Waste Disposal
Fredrick L. Golladay

Examines economic, political, social, and cultural obstacles to meeting the needs of low-income people for water supply and waste disposal. Proposes steps to reduce these obstacles. Notes that communities organize and manage resources to carry out activities that cannot be profitably undertaken by households. Proposes methods to facilitate the formation and operation of community organizations to carry out these functions.

Technical Paper No. 9. 1983. 66 pages.

ISBN 0-8213-0238-8. Stock No. BK 0238. $3.

A Model for the Development of a Self-Help Water Supply Program
Colin Glennie

Presents one version of a practical model for developing, with high community participation, water supply programs in developing countries. Consideration is also given to sanitation program development and practi-

cal guidelines for program development are included. One of a series of informal Working Papers prepared by the Technology Advisory Group, established under UNDP's Global Project, executed by the World Bank.

Technical Paper No. 2. 1982. 45 pages (including 2 annexes, references).

ISBN 0-8213-0077-6. Stock No. BK 0077. $3.

Municipal Water Supply Project Analysis: Case Studies
Frank H. Lamson-Scribner, Jr., and John Huang, editors

Eight case studies and fourteen exercises dealing with the water and wastewater disposal sector.

World Bank (EDI), 1977. 529 pages.

Stock No. IB-0655. $8.50 paperback.

Notes on the Design and Operation of Waste Stabilization Ponds in Warm Climates of Developing Countries
J. P. Arthur

Practical design criteria for pond systems to fit various ambient temperature and reuse requirements, such as irrigation. Offers suggestions for start-up procedures, operation, and trouble shooting.

Technical Paper No. 7. 1983. 106 pages.

ISBN 0-8213-0137-3. Stock No. BK 0137. $5.

Ventilated Improved Pit Latrines: Recent Developments in Zimbabwe
D. D. Mara and Peter R. Morgan

Describes designs and materials for latrines, one for peri-urban areas costing about US$150, and a US$8 version for rural areas. Some 20,000 of the ventilated pit latrines have been built in Zimbabwe. They have been found very effective in eliminating odors and controlling fly breeding.

Technical Paper No. 3. 1983. 48 pages.

ISBN 0-8213-0078-4. Stock No. BK 0078. $3.

Village Water Supply

Describes technical aspects, costs, and institutional problems related to supplying water for domestic use in rural areas and proposes guidelines for future World Bank lending in this sector.

A World Bank Paper. 1976. 98 pages (including 4 annexes).

Stock Nos. BK 9099 (English), BK 9100 (French), BK 9044 (Spanish). $5.

Village Water Supply: Economics and Policy in the Developing World
Robert J. Saunders and Jeremy J. Warford

Addresses the problem of potable water supply and waste disposal in rural areas of developing countries where the majority of the poor tend to be found. Emphasizes the economic, social, financial, and administrative issues that characterize villae water supply and sanitation programs.

The Johns Hopkins University Press, 1976. 292 pages (including 4 appendixes, bibliography, index).

LC 76-11758. ISBN 0-8018-1876-1, Stock No. JH 1876, $22.50 hardcover.

French: L'alimentation en eau des communautes rurales: economie et politique generale dans le monde en developpement. Economica, 1978.

ISBN 2-7178-0022-0, Stock No. IB 0549, $7.

Spanish: Agua para zonas rurales y poblados: economia y politica en el mundo en desarrollo. Editorial Tecnos, 1977

ISBN 84-309-0708-4. Stock No. IB 0532. $7.

Water Supply and Sanitation Project Preparation Handbook
The three companion volumes guide planners, engineers, and community development and public health specialists working in projects related to water and sanitation in developing countries and help users prepare project reports. Case studies illustrate each phase of project development: identification, prefeasibility, and feasibility. Suggests approaches and methods of project evaluation to guide project team members.

Volume 1: **Guidelines**
Brian Grover

Technical Paper No. 12. 1983. 190 pages.

ISBN 0-8213-0230-2. Stock No. BK 0230. $5.

Volume 2: **Case Studies— Identification Report for Port City, Immediate Improvement Project for Port City, Prefeasibility Report for Port City**
Brian Grover, Nicholas Burnett, and Michael McGarry

Technical Paper No. 13. 1983. 352 pages.

ISBN 0-8213-0231-0. Stock No. BK 0231. $15.

Volume 3: Case Study—Feasibility Report for Port City
Brian Grover, Nicholas Burnett, and Michael McGarry

Technical Paper No. 14. 1983. 308 pages

ISBN 0-8213-0232-9. Stock No. BK 0232. $15. Stock No. BK 0367. 3-volume set: $25.

World Bank Studies in Water Supply and Sanitation

The United Nations has designated the 1980s as the International Drinking Water Supply and Sanitation Decade. Its goal is to provide two of the most fundamental human needs—safe water and sanitary disposal of human wastes—to all people. To help usher in this important period of international research and cooperation, the World Bank has published several volumes on appropriate technology for water supply and waste disposal systems in developing countries. Since the technology for supplying water is better understood, the emphasis in these volumes is on sanitation and waste reclamation technologies, their contributions to better health, and how they are affected by water service levels and the ability and willingness of communities to pay for the systems.

Number 1: Appropriate Sanitation Alternatives: A Technical and Economic Appraisal
John M. Kalbermatten, DeAnne S. Julius, and Charles G. Gunnerson

This volume summarizes the technical, economic, environmental, health, and sociocultural findings of the World Bank's research program on appropriate sanitation alternatives and discusses the aspects of program planning that are necessary to implement these findings. It is directed primarily toward planning officials and sector policy advisers for developing countries.

The Johns Hopkins University Press, 1982. 172 pages (including bibliography, index).

LC 80-8963. ISBN 0-8018-2578-4, Stock No. JH 2578, $12.95 paperback.

Number 2: Appropriate Sanitation Alternatives: A Planning and Design Manual
John M. Kalbermatten, DeAnne S. Julius, Charles G. Gunnerson, and D. Duncan Mara

This manual presents the latest field results of the research, summarizes se-lected portions of other publications on sanitation program planning, and describes the engineering details of alternative sanitation technologies and how they can be upgraded.

The Johns Hopkins University Press, 1982. 172 pages (including bibliography, index).

LC 80-8963. ISBN 0-8018-2584-9, Stock No. JH 2584, $15 paperback.

Number 3: Sanitation and Disease: Health Aspects of Wastewater Management
Richard G. Feachem, David J. Bradley, Hemda Garelick, and D. Duncan Mara

Summarizes knowledge about excreta, night soil, and sewage and their effects on health. Describes the environmental characteristics of specific excreted pathogens and the epidemiology and control of the infections these pathogens cause.

John Wiley and Sons (U.K.). 1983. 528 pages.

ISBN 0-471-90094-X. Stock No. IB 0558. $39.50 hardcover.

Appropriate Technology for Water Supply and Sanitation

The following volumes have been published in a less formal format. Some are abridgments of the three foregoing volumes.

Volume 1: Technical and Economic Options
John M. Kalbermatten, DeAnne S. Julius, and Charles G. Gunnerson

Reports technical, economic, health, and social findings of the research project on "appropriate technology" and discusses the program planning necessary to implement technologies available to provide socially and environmentally acceptable low-cost water supply and waste disposal.

1980. 122 pages (including bibliography), Stock No. BK 9186. $5.

Volume 1a: A Summary of Technical and Economic Options
John M. Kalbermatten, DeAnne S. Julius, and Charles G. Gunnerson

A summary of the final report on appropriate technology for water supply and waste disposal in developing countries, a World Bank research project undertaken by the Energy, Water, and Telecommunications Department in 1976-1978.

1980. 38 pages.

Stock No. BK 0049. $3.

Volume 1b: Sanitation Alternative for Low-Income Communities—A Brief Introduction
D. Duncan Mara

Describes in nontechnical language the various low-cost sanitation technologies that are currently available for low-income communities in developing countries and presents a general methodology for low-cost sanitation program planning.

1982. 48 pages.

Stock No. BK 9189. $3.

Volume 3: Health Aspects of Excreta and Sullage Management—A State-of-the-Art Review
Richard G. Feachem, David J. Bradley, Hemda Garelick, and D. Duncan Mara

Provides information on the ways in which particular excreta disposal and reuse technologies affect the survival and dissemination of pathogens. It is intended for planners, engineers, economists, and health workers.

1980. 303 pages (including 14 appendixes, references).

Stock No. BK 9187. $15.

Volume 10: Night-soil Composting
Hillel I. Shuval, Charles G. Gunnerson, and DeAnne S. Julius

Describes a safe, inexpensive treatment method for night-soil composting that is ideally suited for developing countries because of its simplicity in operation, limited need for mechanical equipment, low cost, and effectiveness in inactivating pathogens.

1981. 81 pages (including bibliography, 2 appendixes).

Stock No. BK 9188. $3.

World Bank Research in Water Supply and Sanitation—Summary of Selected Publications

A bibliography summarizing the papers in the Water Supply and Sanitation Series, as well as the World Bank studies in Water Supply and Sanitation published for the World Bank by The Johns Hopkins University Press.

1980.

Stock No. BK 9185. Free.

Prices subject to change without notice and may vary by country.

The World Bank
Publications Order Form

SEND TO: **YOUR LOCAL DISTRIBUTOR** OR TO **WORLD BANK PUBLICATIONS**
(See the other side of this form.)
P.O. BOX 37525
WASHINGTON, D.C. 20013 U.S.A.

Date _____

Name _____

Title _____

Firm _____

Address _____

City _____ State _____ Postal Code _____

Country _____ Telephone (_____) _____

Purchaser Reference No. _____

Ship to: (Enter if different from purchaser)

Name _____

Title _____

Firm _____

Address _____

City _____ State _____ Postal Code _____

Country _____ Telephone (_____) _____

Check your method of payment.
Enclosed is my ☐ Check ☐ International Money Order ☐ Unesco Coupons ☐ International Postal Coupon.
Make payable to World Bank Publications for U.S. dollars unless you are ordering from your local distributor.

Charge my ☐ VISA ☐ MasterCard ☐ American Express ☐ Choice. (Credit cards accepted only for orders addressed to World Bank Publications.)

_____ _____ _____
Credit Card Account Number   Expiration Date   Signature

☐ Invoice me and please reference my Purchase Order No. _____.

Please ship me the items listed below.

Stock Number	Author/ Title	Customer Internal Routing Code	Quantity	Unit Price	Total Amount $

All prices subject to change. Prices may vary by country. Allow 6-8 weeks for delivery.

Subtotal Cost $_____

Total copies _____ Air mail surcharge if desired ($2.00 each) $_____

Postage and handling for more than two complimentary items ($2.00 each) $_____

Total $_____

Thank you for your order.

IBRD-0053

Distributors of World Bank Publications

ARGENTINA
Carlos Hirsch, SRL,
Attn: Ms. Monica Bustos
Florida 165 4° piso
Galeria Guemes
Buenos Aires 1307

AUSTRALIA, PAPUA NEW GUINEA,
FIJI, SOLOMON ISLANDS,
WESTERN SAMOA, AND
VANUATU
The Australian Financial Review
Information Service (AFRIS)
Attn: Mr. David Jamieson
235-243 Jones Street
Broadway
Sydney, NSW 20001

BELGIUM
Publications des Nations Unies
Attn: Mr. Jean de Lannoy
av. du Roi 202
1060 Brussels

CANADA
Le Diffuseur
Attn: Mrs. Suzanne Vermette
C.P. 85, Boucherville J4B 5E6
Quebec

COSTA RICA
Libreria Trejos
Attn: Mr. Hugo Chamberlain
Calle 11-13, Av. Fernandez Guell
San Jose

DENMARK
Sanfundslitteratur
Attn: Mr. Wilfried Roloff
Rosenderns Alle 11
DK-1970 Copenhagen V.

EGYPT, Arab Republic of
Al Ahram
Attn: Mr. Sayed El-Gabri
Al Galaa Street
Cairo

FINLAND
Akateeminen Kirjakauppa
Attn: Mr. Kari Litmanen
Keskuskatu 1, SF-00100
Helsinki 10

FRANCE
World Bank Publications
66, avenue d'Iéna
75116 Paris

GERMANY, Federal Republic of
UNO-Verlag
Attn: Mr. Joachim Krause
Simrockstrasse 23
D-5300 Bonn 1

HONG KONG, MACAU
Asia 2000 Ltd.
Attn: Ms. Gretchen Wearing Smith
6 Fl., 146 Prince Edward Road
Kowloon

INDIA
UBS Publishers' Distributors Ltd.
Attn: Mr. D.P. Veer
5 Ansari Road, Post Box 7015
New Delhi 110002
(Branch offices in Bombay, Bangalore,
Kanpur, Calcutta, and Madras)

INDONESIA
Pt. Indira Limited
Attn: Mr. Bambang Wahyudi
Jl, Dr. Sam Ratulangi No. 37
Jakarta Pusat

IRELAND
TDC Publishers
Attn: Mr. James Booth
12 North Frederick Street
Dublin 1

JAPAN
Eastern Book Service
Attn: Mr. Terumasa Hirano
37-3, Hongo 3-Chome, Bunkyo-ku 113
Tokyo

KENYA
Africa Book Services (E.A.) Ltd.
Attn: Mr. M.B. Dar
P.O. Box 45245
Nairobi

KOREA, REPUBLIC OF
Pan Korea Book Corporation
Attn: Mr. Yoon-Sun Kim
P.O. Box 101, Kwanghwamun
Seoul

MALAYSIA
University of Malaya Cooperative
Bookshop Ltd.
Attn: Mr. Mohammed Fahim Htj
Yacob
P.O. Box 1127, Jalan Pantai Baru
Kuala Lumpur

MEXICO
INFOTEC
Attn: Mr. Jorge Cepeda
San Lorenzo 153-11, Col. del Valle,
Deleg. Benito Juarez
03100 Mexico, D.F.

NETHERLANDS
MBE BV
Attn: Mr. Gerhard van Bussell
Noorderwal 38,
7241 BL Lochem

NORWAY
Johan Grundt Tanum A.S.
Attn: Ms. Randi Mikkelborg
P.O. Box 1177 Sentrum
Oslo 1

PANAMA
Ediciones Libreria Cultural Panamena
Attn: Mr. Luis Fernandez Fraguela R.
Av. 7, Espana 16
Panama Zone 1

PHILIPPINES
National Book Store
Attn: Mrs. Socorro C. Ramos
701 Rizal Avenue
Manila

SAUDI ARABIA
Jarir Book Store
Attn: Mr. Akram Al-Agil
P.O. Box 3196
Riyadh

SINGAPORE, TAIWAN, BURMA
Information Publications Private, Ltd.
Attn: Ms. Janet David
02-06 1st Floor, Pei-Fu Industrial
    Building
24 New Industrial Road
Singapore

SPAIN
Mundi-Prensa Libros, S.A.

Attn: Mr. J.M. Hernandez
Castello 37
Madrid

SRI LANKA AND THE MALDIVES
Lake House Bookshop
Attn: Mr. Victor Walatara
41 Wad Ramanayake Mawatha
Colombo 2

SWEDEN
ABCE Fritzes Kungl, Hovbokhandel
Attn: Mr. Eide Segerback
Regeringsgatan 12, Box 16356
S-103 27 Stockholm

SWITZERLAND
Librairie Payot
Attn: Mr. Henri de Perrot
6, rue Grenus
1211 Geneva

TANZANIA
Oxford University Press
Attn: Mr. Anthony Theobold
Maktaba Road, P.O. Box 5299
Dar es Salaam

THAILAND
Central Department Store, Head Office
Attn: Mrs. Ratana
306 Silom Road
Bangkok

Thailand Management Association
Attn: Mrs. Sunan
308 Silom Road
Bangkok

UNITED KINGDOM AND
NORTHERN IRELAND
Microinfo Ltd.
Attn: Mr. Roy Selwyn
Newman Lane, P.O. Box 3
Alton, Hampshire GU34 2PG
England

UNITED STATES
The World Bank Book Store
600 19th Street, N.W.
Washington, D.C. 20433
(Postal address: P.O. Box 37525
Washington, D.C. 20013, U.S.A.)

Baker and Taylor Company
501 South Gladiola Avenue
Momence, Illinois, 60954

380 Edison Way
Reno, Nevada, 89564

50 Kirby Avenue
Somerville, New Jersey, 08876

Commerce, Georgia 30599

Bernan Associates
9730-E George Palmer Highway
Lanham, Maryland, 20761

Blackwell North America, Inc.
1001 Fries Mill Road
Blackwood, New Jersey 08012

Sidney Kramer Books
1722 H Street, N.W.
Washington, D.C. 20006

United Nations Bookshop
United Nations Plaza
New York, N.Y. 10017

VENEZUELA
Libreria del Este
Attn. Mr. Juan Pericas
Avda Francisco de Miranda, no. 52
Edificio Galipan, Aptdo. 60.337
Caracas 1060-A